FOR
Evan G. Davis and Gene P. Davis

THE POLITICS OF HAZARDOUS WASTE

THE POLITICS OF HAZARDOUS WASTE

CHARLES E. DAVIS
Colorado State University

PRENTICE HALL, Englewood Cliffs, New Jersey 07632

y of Congress Cataloging-in-Publication Data

3, CHARLES E.
he politics of hazardous waste/Charles E. Davis.
p. cm.
ncludes bibliographical references and index.
SBN 0-13-683202-4
. Hazardous wastes—Government policy—United States. I. Title.
1040.D38 1993
.72'8756'0973—dc20 92-29995
CIP

uisitions editor: Julie Berrisford
rior design and electronic page makeup: Rob DeGeorge
editor: Eleanor Walter
ress buyer: Kelly Behr
ufacturing buyer: Mary Ann Gloriande
Cover design: Patricia Kelly
Editorial assistant: Nicole Signoretti

A Simon & Schuster Company
Englewood Cliffs, NJ 07632

Printed in the United States of America

10 9 8 7 6 5 4 3 2 1

ISBN 0-13-683202-4

PRENTICE-HALL INTERNATIONAL (UK) LIMITED, *London*
PRENTICE-HALL OF AUSTRALIA PTY. LIMITED, *Sydney*
PRENTICE-HALL CANADA INC., *Toronto*
PRENTICE-HALL HISPANOAMERICANA, S.A., *Mexico*
PRENTICE-HALL OF INDIA PRIVATE LIMITED, *New Delhi*
PRENTICE-HALL OF JAPAN, INC., *Tokyo*
SIMON & SCHUSTER ASIA PTE. LTD., *Singapore*
EDITORA PRENTICE-HALL DO BRASIL, LTDA., *Rio de Janeiro*

CONTENTS

PREFACE

"Chemical Wastes: A Buried Bombshell."* Headlines of this sort from a decade ago illustrate an important truth about hazardous waste policy issues. In contrast to the atmosphere surrounding the Clean Air Act of 1970, which was enacted during an era of heightened environmental consciousness, the reaction of elected officials to waste disposal problems has been more appropriately described as the politics of fear. Many people in the United States have become aware of toxic waste through powerful and occasionally exaggerated images of ecological devastation transmitted through media coverage of Love Canal and other tragedies. Not surprisingly, attempts to come up with policy solutions have been influenced by the need to address public concerns, despite a lack of policy precedent or a well-developed body of information on the control, reuse, or eradication of wastes. Political criteria have clearly assumed a more prominent role in the hazardous waste policymaking process than economic or technical factors.

In this book, I will focus on the manner in which political institutions and public officials deal with controversial and emotional U.S. policy issues such as hazardous waste. At one level, the text offers a single-issue example of how the policymaking process works. The scope of

**U.S. News and World Report*, September 29, 1980, p. 39.

waste disposal problems is discussed in the introductory chapter. Here the notion of problem definition and its attendant political consequences for the regulated community is fleshed out. Next, attention is directed to the enactment and implementation of the two principal federal statutes affecting the management and cleanup of hazardous wastes; i.e., the Resource Conservation and Recovery Act (RCRA) and the Comprehensive Environmental Response, Compensation, and Liability Act (better known as Superfund). The concluding chapter is devoted to an examination of how well (or poorly) political institutions have performed in meeting their decision-making responsibilities and a brief discussion of several important policy issues requiring the attention of policymakers.

I have also attempted to add a few new wrinkles to the text on matters that have rarely received their due in policy process books. The importance of policy learning is keenly reflected in the statutory reauthorization of both RCRA and Superfund, a point that may be true for other policy areas hobbled by technological uncertainty. In addition, the role played by state and local political institutions is highlighted here. Nonfederal governments are at the front lines of community controversies dealing with the siting of hazardous waste facilities and the implementation of right-to-know laws. Not only do their actions have a direct bearing on the likelihood of success in a programmatic sense, but they also remind us of the high degree of interdependence between the two laws. Facilities that fail to carry out good environmental management practices today are prime candidates for future listing as state or federal Superfund sites.

As usual, the efforts reflected in the following pages are a product of multiple sources of information, advice, and encouragement. Jim Lester has provided considerable help as a colleague and occasional collaborator. Not only has he shared what he knows about the subject (which is a great deal), but he has acted as a useful sounding board as well. Others within the small community of hazardous waste researchers have contributed through their writings or in unwritten insights offered at professional meetings. Julie Berrisford at Prentice Hall gave encouragement early on to tackle this subject in book-length form. The following reviewers also helped me to shape the text into its present form: W. Douglas Costain, University of Colorado at Boulder; Susan Hunter, West Virginia University; Daniel McCool, University of Utah; William Mangun, East Carolina University; Zachary Smith, Northern Arizona University; and Richard Tobin, State University of New York at Buffalo. Finally, I would like to acknowledge the patience and advice of my wife, Sandy, who has been supportive throughout the process of putting this project together.

C.E.D.
Fort Collins, Colorado

THE POLITICS OF HAZARDOUS WASTE

HAZARDOUS WASTE AS A POLICY PROBLEM

The problem of hazardous waste contamination in the United States presents us with a political paradox of sorts. On the one hand, the pollsters suggest that this issue is more likely to provoke citizen anxiety about health-related risks than problems related to air or water pollution. Clearly, hazardous waste is an issue that has received considerable attention from the media, industry and environmental advocacy groups, government officials, and, on occasion, the general public. On the other hand, scientists are not inclined to share the public view that hazardous waste is among the most serious environmental problems confronting policy makers. A recent study by the U.S. Environmental Protection Agency's Science Advisory Board has concluded that the mismanagement of toxic waste is less likely to harm human health than is inattention to the control of other pollutants.[1]

What should be done? Coming up with a politically acceptable means of managing hazardous waste has proven to be enormously frustrating for elected officials. Tension arises in the political process because of the difficulty of reconciling diverse and sometimes competing concerns. Resolution of these concerns requires a greater appreciation for the interplay between science, economics, and politics than is evident for other areas of policy. The complexity of hazardous waste policy issues thus provides a test for American political institutions at all levels of government and at differing stages of the policymaking process.

The primary objective of this book is to shed light on the formulation, adoption, and implementation of hazardous waste laws in the United States. Attention is restricted to a pair of key statutes. In 1976, Congress adopted the Resource Conservation and Recovery Act (RCRA), which established a regulatory framework for the management of toxic chemical wastes from the point of generation to their ultimate storage, disposal, or recovery. RCRA was amended in 1984 and regulatory coverage was expanded to embrace a larger number of organizations.

A second policy, the Comprehensive Environmental Response, Compensation, and Liability Act (better known as Superfund), was enacted in 1980 and provided funding for the cleanup of abandoned hazardous waste dump sites. It was reauthorized in 1986 with a substantial budgetary increase. Both policies are administered by the federal Environmental Protection Agency (EPA) and call for varying degrees of state participation in the enforcement of regulatory standards.

A secondary objective of this text is to assess the contributions of political institutions, such as Congress, EPA, and state environmental agencies, in the hazardous waste policymaking process. Are these institutions able and willing to make policy decisions aimed at reducing the risks of chemical contaminants? Or are they likely to encounter obstacles in the form of technical, economic, and political constraints that limit the programmatic effectiveness of RCRA and Superfund?

DEFINING HAZARDOUS WASTE

To say that a hazardous waste problem exists implies that we know what it is. EPA has defined *hazardous waste* in two different ways under authority granted by RCRA. First, a chemical may be considered hazardous if it exhibits certain characteristics; i.e., ignitability, reactivity, corrosivity, or toxicity. Second, a number of hazardous wastes have been listed by name. Both industrial wastes and selected commercial chemical products have been included in EPA's inventory.

Despite efforts by EPA officials to provide a definition that is workable for program management purposes, the existing list of hazardous chemicals is by no means exhaustive or all-encompassing. Some wastes are not included under the RCRA umbrella because of regulatory coverage elsewhere. For example, pesticides fall within the program jurisdiction of the Federal Insecticide, Fungicide, and Rodenticide Act. Other pollutants are regulated by EPA (under the Clean Water Act, the Safe Water Drinking Act, the Toxic Substances Control Act, and the

Clean Air Act), the Department of the Interior (under the Surface Mining Control and Reclamation Act), and the Department of Transportation (under the Hazardous Materials Transportation Act).[2]

A second factor that accounts for a less extensive list of hazardous wastes is exemption by statute. In the 1980 Amendments to the Solid Waste Disposal Act, Congress specifically removed several types of industry waste from regulatory coverage even though these substances were thought to contain hazardous characteristics. Included were wastes associated with the extraction and processing of minerals, fly ash produced as a byproduct of coal combustion, and drilling muds used in oil field operations. Special exemptions were also granted for wastes targeted for recycling or for use as a fuel source.

A third reason for not incorporating additional wastes under RCRA is, quite simply, a lack of consensus between EPA, other federal agencies, and state environmental departments over what constitutes a "hazardous waste." One of the major sources of disagreement has been the EPA standard for toxicity.[3] Agency officials were confronted with a statutory provision placing the burden of proof on EPA for the listing of new hazardous wastes. Each could be included only after extensive testing and analysis. Thus, a narrower conception of toxicity was developed because of the time, expense, and staffing implications of a more inclusive definition. According to Greer, a more efficient strategy for EPA officials was to place more emphasis on a waste characteristics approach than a listing approach, since:

> Identifying waste by characteristic...does not require EPA documentation on any particular waste; rather, the Agency need only show that any waste which demonstrates this characteristic would pose a hazard when mismanaged. Identification of hazardous waste by characteristic thus has a greater potential for broadening the scope of EPA coverage, and the use of characteristics to bring wastes into the RCRA system has the broadest potential for impact.[4]

Dissatisfaction with the EPA toxicity standard led to the development of a more comprehensive definition of hazardous waste in several states. For example, California has classified many household, agricultural, and mining wastes; drilling muds; and sewage sludge as hazardous in addition to the wastes listed under RCRA.[5] Congress also began to display concern about the small number of wastes registered by EPA and the glacial pace at which new wastes were added. In the 1984 Hazardous and Solid Waste Amendments, EPA was directed to expand its toxicity characteristic.[6] Sponsors of this directive clearly intended that its implementation would result in a sizable increase in the number of wastes regulated under RCRA as well as greater compatibility between federal and state lists.

THE SCOPE OF THE PROBLEM

Having considered what we mean by hazardous waste, we must examine the scope of the problem. Of particular concern are the amount and type of wastes generated; industry and governmental sources of waste production; the geographical distribution of toxic chemical substances; and management approaches used to control, eliminate, or recover these wastes. A study conducted by the Congressional Budget Office (CBO) concluded that approximately 266 million metric tons of hazardous waste are generated in the United States annually, which amounts to more than one ton per person residing in the country.[7] Put differently, the yearly volume of American waste is sufficiently large to cover the island of Manhattan under a single-story pile.[8]

The production of hazardous waste as an unwanted industrial byproduct is not a terribly new phenomenon. However, the rate of hazardous waste generation has steadily increased since the end of World War II, when the volume of chemicals and accompanying wastes generated was relatively low. Much of this increase can be attributed to growth in petroleum-based chemical products such as pesticides, synthetic fabrics, plastics, wood preservatives, new paints, and solvents, among others.[9]

Who are the primary generators of hazardous waste? EPA has begun collecting data summarizing the amounts of waste released directly onto the land or underground as well as wastes transported off-site to treatment, storage, or disposal (tsd) facilities by waste-generating organizations. The most recent figures indicate that the chemical products industry alone accounts for nearly half of the U.S. hazardous waste total, while firms involved in the manufacture or processing of primary metals are responsible for approximately one-seventh. Other important sources of hazardous waste include the petroleum, paper, food, and fabricated metals industries.[10]

One other producer of hazardous wastes is worth mentioning—the U.S. government. Like everyone else, executive departments and agencies must comply with the law. However, in a collective sense, federal bureaucracies have not set a good example for industry or other regulated parties in choosing responsible environmental management practices. A study by the Congressional Budget Office revealed the existence of 1,099 contaminated sites identified by various agencies and an additional 8,357 sites identified by researchers in need of remedial work. Overall, federal facilities constitute approximately ten percent of the seriously polluted sites found on EPA's National Priorities List.[11]

Sources of contamination within the federal government are quite diverse. They include research laboratories within the Agriculture and Health and Human Services Departments, abandoned mines on public lands, toxic byproducts of illegal drugs obtained in raids carried out by the Drug Enforcement Administration, nuclear weapons plants, military bases, and numerous properties acquired by federal financial regulators following the seizure of assets from failed savings and loan institutions. However, some parties clearly contribute more than others to the accumulation of waste on federal property. The Department of Defense (DOD) is responsible for the largest number of contaminated sites, while the most serious pollution problems can be found in nuclear weapons facilities managed by the Department of Energy (DOE).[12]

How are toxic chemical releases onto the land or underground distributed geographically? An analysis of EPA's Toxic Release Inventory (TRI) data indicates that a relatively small number of states are responsible for a disproportionately large percentage of the national totals. Within the land disposal category, the leading states are Florida and Indiana, releasing 82 million pounds and 63 million pounds, respectively. Other states contributing a large amount of waste to land-based facilities include Arizona, Missouri, Montana, Ohio, and Texas.[13]

A regional breakdown reveals no dominant area, although states located in the South, the industrial Midwest, the Atlantic Coast, and the Rocky Mountains tend to rank higher in terms of the amount of chemical wastes disposed of on the land. Least likely to release wastes onto land-based facilities are states located in New England, the Great Plains, and the Pacific Northwest. While these figures indicate a rather wide geographical dispersal of wastes ending up in repositories, the opposite is true for the underground injection of hazardous chemical substances. Fully two-thirds of all wastes disposed of in this fashion nationally are attributable to industrial operations in Texas and Louisiana.[14]

A statistical profile of this sort can communicate information about the "who," "how much," and "where" of waste production to policymakers, but it says little about attendant risks to public health or environmental quality. Also worth considering are industry and governmental efforts to control hazardous wastes. For example, what is the proportion of hazardous waste managed on-site (on the premises of waste-generating firms) in relation to wastes sent to an off-site location (a treatment, storage, and disposal facility)? What approaches to the treatment, disposal, or recovery of wastes are most commonly used by industry? Do environmental administrators have information about hazardous waste generators, transporters, and tsd facilities that is sufficiently broad, reliable, and up-to-date for them to make good decisions?

Despite efforts by federal and state policymakers to move away from land-based approaches to hazardous waste management, most wastes continue to be managed on the ground. One of the most widely used methods of disposal has been the injection of wastes into wells or salt caverns, although this practice has declined since the adoption of the 1984 RCRA amendments. This technique has been popular with industry because of ample storage capacity, low disposal costs, and the relative ease of gaining a permit from governmental agencies. It also has generated relatively little media attention or citizen involvement. In contrast, attempts to construct aboveground facilities for waste treatment, storage, or disposal have often been thwarted by public opposition from nearby communities.

Other approaches commonly used by pollution-generating firms include the discharge of treated and untreated wastes into municipal sewage treatment plants, rivers, and streams; surface impoundments (placing liquid wastes and sludge into pits, ponds, or lagoons); and landfills. However, reliance on these methods can create problems. For example, an EPA study concluded that 70 percent of the 80,000 surface impoundments accepting hazardous wastes are constructed without liners. Agency scientists estimated the risk of groundwater contamination to be quite high, around 90 percent.[15]

Will greater reliance on appropriate technology solve the problem? Not necessarily. Even landfills operating with the prescribed plastic liners designed to prevent wastes from leaking into the soil are considered to be less than satisfactory since few of the wastes have been pretreated to enhance rapid decomposition prior to disposal. There are also several approaches that give greater emphasis to waste treatment, source reduction, or recycling than land-based management options, but they do not yet account for a significant share of the waste-disposal market because of cost considerations.

Another concern associated with the management of toxic chemical substances is the proportion of wastes handled on-site versus off-site. The vast majority of wastes (96 percent) are stored or disposed of on the premises of pollution-generating companies. Many of these companies generate a large volume of waste, thereby increasing their financial incentive to invest in on-site or nearby disposal facilities while minimizing transportation costs. This is less true for smaller firms and certain types of firms (such as manufacturers of rubber and plastics products) that find it "more cost-effective to send wastes to larger commercial waste management facilities with greater economies of scale."[16]

The effective implementation of RCRA and Superfund by EPA and state environmental agencies requires information that is comprehensive, reliable, and up-to-date. These statutes call for the collection of data to be used for regulatory decision-making purposes. Responsibilities for acquiring and processing this information are shared by the federal government, state environmental agencies, and industry. For example, EPA initiates and assembles research on hazardous waste and maintains a list of priority sites for cleanup actions authorized under Superfund.

Industries are required to keep records of the type and quantity of wastes generated and must prepare transportation manifests if these wastes are sent to a tsd facility. State environmental agencies maintain a list of wastes and waste handlers involved in the generation, transport, recycling, or disposal of waste; these lists are often more extensive than the list kept by EPA. They also participate in the collection and analysis of data culled from the transport manifests.

The adequacy of the database clearly has consequences for the success or failure of program enforcement efforts. Accordingly, concerns have been expressed about the usefulness of existing sources of information. In its 1983 report, the Office of Technology Assessment (OTA) indicated that the EPA list of hazardous waste had little management value since the actual universe of wastes posing a threat to public health or the environment was considerably larger. Its primary use lay in providing an information base for the allocation of federal grant monies to state environmental agencies.[17]

A second area of concern has been the completeness and validity of data submitted to environmental agencies by regulated industries. A study by the General Accounting Office concluded that administrators cannot make reasonable estimates about groundwater quality at many land disposal facilities because facility owners and operators have not fully complied with information reporting requirements.[18]

In short, the amount of hazardous waste generated by industry and government is both significant and on the rise, but existing efforts to manage these wastes have not served to appreciably reduce risks to public health or the environment.[19] The development of policy solutions has been limited to some extent by the lack of information on waste-related impacts. But there are additional complications stemming from interest group or agency priorities that are occasionally at odds, notably the desire for decision-making options with greater economic efficiency and/or political support. Each is discussed below.

Technical Uncertainty

A lack of information about hazardous waste or disagreement over attributes or effects of toxic chemical substances can seriously erode efforts to develop or implement policy. Examples of how technological uncertainty affects decision making include efforts by EPA to promulgate regulations and to produce a ranking system for estimating risks to human health or the environment from exposure to a given hazardous waste site. Issuing rules under the authority of RCRA has been complicated by the need to meet deadlines established by Congress without sufficient information or resources. This, in turn, has led to delay or the development of hastily conceived regulation with unintended consequences. One such "regulatory surprise" is discussed by Rosenbaum:

> EPA's regulations for RCRA exclude from control any combustible liquid wastes with a "beneficial reuse" as boiler fuel. In fact, a much larger volume of industrial hazardous waste than had been supposed is combined with petroleum or other combustible materials and burned in boilers—in effect, grandfathered out of the law. Unfortunately, the furnaces in which these waste-spiked fuels are burned usually are not designed to control the hazardous chemicals in the fuel and consequently many hazardous substances are released into the atmosphere contrary to the law's intent.[20]

While the prospect of rule making aided by imperfect information is decidedly unappealing, a decision to delay the process can be politically unwise. In 1978, EPA was sued by the Environmental Defense Fund for failing to meet congressional deadlines on the development of rules for RCRA. Subsequently, legislative frustration with EPA's inability to produce program rules in an expeditious manner led to the inclusion of a "hammer clause" in the Hazardous and Solid Waste Amendments of 1984: If EPA fails to act within the timetable established by Congress, environmental standards created elsewhere (the state of California) become operative.[21]

Estimates dealing with the magnitude of hazardous waste problems in the United States or a given political jurisdiction also can be challenged because of disagreement over the methods of analysis used to reach a set of conclusions. One such example is the hazardous waste ranking system adopted by EPA to help state governments inventory and evaluate hazardous waste sites nationwide. The importance of this process is linked with the assignment of rank or risk potential, since this determines whether a site is included on the National Priorities List (NPL), and thereby obtaining eligibility for Superfund monies. By the end of 1984, nearly 20,000 sites had been analyzed and incorporated within a database referred to as the Emergency and Remedial Response Information System (ERRIS).[22]

However, there is reason to believe that the kinds of assumptions used in the ranking system are sufficiently strict to exclude from consideration heavily polluted sites that would ordinarily qualify for inclusion on the NPL. Major criteria used in this process include the population at risk; a state's ability to absorb certain costs and responsibilities; and the potential for a site to contaminate drinking water, produce direct human contact, or destroy sensitive ecosystems. EPA adopted a decision rule calling for the acquisition of complete information on any given site prior to evaluation; otherwise the analysis would not be performed. A case study focusing on Florida's experience with this process revealed that only 27 of the 160 to 200 sites identified as dangerous by the state's Department of Environmental Regulation were actually examined by EPA.[23]

Economic Efficiency

Environmental administrators and industry officials can occasionally agree on hazardous waste policies or management approaches on the basis of technical criteria but may find themselves at odds over specific regulatory requirements because of compliance costs. One source of tension has been attributed to a structural feature of RCRA commonly found in many environmental statutes. Uniform standards for environmental quality have been established by EPA, and they affect firms differently on the criteria of size, region, and technology. Meeting hazardous waste statutory requirements places a much greater financial burden on smaller firms, since their compliance costs in relation to profit margin are higher than those of larger firms. In addition, the costs of off-site disposal vary by region. Waste-generating companies operating in the industrial states of the Northeast and the Midwest are likely to find tsd facilities within easy driving distance. Similar firms headquartered in the Rocky Mountain states are more apt to be confronted with high transportation costs, since available facilities tend to be farther away.

Third, between-firm differences in production facilities ensure that some companies will need to expend more resources than others to comply with RCRA regulations. Some firms can get by with few process changes, while others find it necessary to underwrite significant alterations in production technology to produce less waste. These factors can adversely affect a firm's competitiveness in the marketplace and may well increase the temptation to dispose of wastes cheaply—and illegally.[24]

It is also reasonable to ask whether the architects of RCRA devoted sufficient attention to incentives built into their regulatory design for the selection of waste-disposal options. There is general agreement that source reduction or producing less waste through change in methods of production, treatment, and recycling is an environmentally sound approach, while land-based containment approaches are less desirable; however, as Table 1–1 suggests, an inverse relationship is found between risk and cost.

There are several reasons why on-site containment options continue to be popular with waste-generating firms. First, transportation costs are minimized or eliminated. Second, the administrative costs associated with preparing the manifest for off-site waste disposal are avoided, as are fees charged by tsd facility operators. Third, industry

TABLE 1–1 Estimated Costs of Hazardous Waste Treatment and Disposal Methods

TREATMENT AND DISPOSAL METHODS	COST PER TON (CONSTANT DOLLARS)
Publicly owned treatment works	$140–$350
Solid waste landfill (bulk)	under $50
Hazardous waste landfill (drummed)	$160–$240
Hazardous waste landfill (with stabilization)	$75–$175
Land treatment	under $50
Deepwell injection (oily rinse)	under $50
Deepwell injection (toxic rinse)	$125–$260
Chemical treatment (acids and alkalines)	$25–$90
Chemical treatment (heavy metals, cyanide)	$50–$750
Solvent recycling	$0–$425
Incineration (liquids)	$0–$225
Incineration (solid, highly toxic)	$375–$750
Incineration (drummed)	$100–$1,000
Incineration (municipal waste to energy)	under $50

Source: Conservation Foundation, *State of the Environment: A View toward the Nineties* (Washington, D.C.: Conservation Foundation, 1987).

officials are aware that surface impoundments or lagoons are unlikely to be inspected and/or cited for improper management practices in a timely fashion. A study by the General Accounting Office concluded that RCRA enforcement actions have not received sufficient budgetary or staff resources to provide a credible regulatory deterrent.[25] Over the long run, this problem may be lessened by policy and administrative changes mandated by the Hazardous and Solid Waste Amendments of 1984, which are designed to produce a better fit between the economics of waste disposal and the selection of abatement or preventive approaches that are environmentally benign.

Political Support

Hazardous waste policy problems can become complicated by political as well as economic or technical factors. A key concern is the presence or absence of program commitment on the part of elected officials and administrators with important policymaking responsibilities. For example, high-level EPA officials were noticeably lukewarm in their support for the policy goals of RCRA and Superfund between 1981 and 1983. EPA Administrator Anne Gorsuch (later Burford) and Rita Lavelle, an assistant administrator in charge of hazardous waste programs, served notice early on that priority would be given to Reagan Administration policies such as the reduction of inflation, a lessening of the regulatory burden on the private sector, and efforts to develop more cooperative relationships with industry.[26]

A lesser commitment to hazardous waste programs in relation to larger economic development concerns was reflected in administrative decisions such as an unsuccessful attempt to delete the ban on the disposal of liquid chemical wastes into landfills and cuts in the RCRA budget. Because of these and related actions, a message was effectively communicated to industry officials—that breaking the laws would not result in severe punishment.[27] In addition, political problems at the top of the policymaking hierarchy produced unwanted ripple effects: notably the less effective use of financial and staff resources in agency decision making and a decline in employee morale.

The importance of political support as a decision-making criterion is not restricted to public-sector officials. Citizens have traditionally been strongly opposed to the prospective siting of a hazardous waste tsd facility in or near their community. Quite apart from the actual risk posed by facility construction and operation is the *perception* of risk by local residents. Environmental problems are more easily accepted by the public if they are viewed as "voluntary, controllable, known, familiar, and immediate." [28] However, it is difficult to apply these attributes to siting controversies. Air or water pollution problems cannot generate the same intensity of public concern as hazardous waste problems, since they are often viewed as resolvable if sufficient political and financial resources are allocated to the task at hand.[29] The political consequence of fear combined with uncertainty is the so-called NIMBY (not-in-my-backyard) syndrome, which has made it extremely difficult for would-be developers to find a governmental jurisdiction willing to house their tsd facility. This syndrome has also served to exacerbate the problem of unsafe management practices, since pollution-generating firms have an incentive to dispose of wastes on-site rather than off-site.

The relative importance of hazardous waste policy criteria tends to vary according to institutional concerns, level of government, and stage of the policymaking process. EPA is clearly in a delicate position because of the need to balance technical and political factors in reaching a decision. Technical proficiency is a critical prerequisite for the development of regulations with a sufficiently sound evidentiary basis to survive administrative scrutiny from the federal Office of Management and Budget and legal challenges from affected industries.[30]

Congress is especially attentive to the concerns of key political constituencies on high-visibility issues. This is illustrated by the legislative battle over the adoption of Superfund in 1980 and the decision by the chairmen of six congressional committees to hold hearings in the wake of political scandals involving Superfund contracts and high-level EPA officials in 1983. Political factors also are important to EPA officials within the context of their role in the formulation and implementation of hazardous waste programs.

The policy actor most concerned about economic efficiency of hazardous waste programs at the federal level is the Office of Management and Budget (OMB). Its role in the evaluation of program budget requests has traditionally been carried out with an eye toward fiscal prudence. In 1981, OMB was given a larger responsibility in the evaluation of rules proposed by executive agencies as well, requiring that benefits to society exceed the costs of industry compliance (these changes are discussed more fully in Chapter 3). State and local governmental agencies have demonstrated considerable concern about economic criteria as well but for differing reasons. Subnational governments are occasionally placed in the position of having to weigh the impact of stricter hazardous waste programs against the possibility that increased compliance costs may lead industry officials to consider relocating to another state.

THE REST OF THE BOOK

A preliminary look at the difficulties associated with defining hazardous waste and determining the magnitude of pollution problems that accompany unsafe or illegal disposal practices gives us a fairly good hint that efforts to obtain a political solution are neither quick nor easy. Chapter 2 provides an analysis of how policies are formulated. Attention is directed to the rather disparate routes to the governmental agenda taken by RCRA and Superfund and the politics of building the necessary support in Congress for the adoption and reauthorization of these policies.

Next, the political context of the policymaking process is discussed in Chapter 3. EPA must make decisions in a political and administrative fishbowl. Agency administrators not only must perform in a manner that satisfies both Congress and their hierarchical superiors in the White House but must contend with an array of additional political and institutional hurdles as well. The executive branch is laden with political landmines, including the Office of Management and Budget, which oversees proposed rules and budgets and the rather onerous task of getting other departments to comply with hazardous waste laws. In addition, the statutory, organizational, and resource contexts also constrain or in some cases contribute to EPA's decision-making capabilities.

Chapter 4 describes EPA's performance in carrying out critical responsibilities such as the communication of policy objectives, the delegation of program management authority to the states, regulation (especially rule making and the issuance of permits), oversight of state-administered RCRA and Superfund programs, and enforcement. While inspections, administrative orders, and lawsuits represent the most direct form of implementation, the other activities are designed to shape the behavior of regulated parties through the manipulation of legal and economic incentives.

The central purpose of Chapter 5 is to assess the other key set of implementors—state and local officials. We seek to determine whether the states have the political, financial, and organizational wherewithal to handle hazardous waste programs. On the one hand, public officials have considerable leeway to tailor policies in ways that fit within state needs after minimum federal standards for environmental quality have been met. On the other hand, these officials must grapple with both resource constraints and political pressures from regulated industries to minimize compliance costs, with the not-so-subtle message that actions (such as the promulgation of strict regulations) have consequences (such as the departure of affected industries to states with a more favorable business climate).

In many respects the "mandate millstone," a term coined by former New York City Mayor Ed Koch to describe the growing financial burden imposed on local government through regulatory statutes enacted by federal and state officials, is especially applicable to hazardous waste program implementation at the local level. Coming to grips with such contentious political issues as what to do with leaking underground storage tanks for petroleum and siting a tsd facility often falls on the shoulders of community officials. They may ultimately bear a disproportionate share of the political and economic costs for addressing problems at the stage of the policymaking process where perceived risk is greatest.

The final chapter begins with an assessment of how well political institutions with important responsibilities for hazardous waste policy formulation or implementation have performed within the limits of available resources. Next, ongoing policy issues that are likely to complicate the subsequent reauthorization of RCRA or Superfund are examined. We end with a brief discussion of hazardous waste as a policy issue. In what ways is it similar to or unlike other substantive policy areas?

NOTES

1. U.S. Environmental Protection Agency, Science Advisory Board, *Reducing Risk: Setting Priorities and Strategies for Environmental Protection* (Washington, D.C.: Government Printing Office, September 1990), especially Appendix C.

2. For a useful description of assorted pollution-control programs and their interrelationships, consult Office of Technology Assessment, *Technologies and Management Strategies for Hazardous Waste Control* (Washington, D.C.: Government Printing Office, 1983), pp. 319–27 (hereafter referred to as the OTA Report).

3. Linda E. Greer, "Definition of Hazardous Waste," *Hazardous Waste*, 1, No. 3 (1984), pp. 309–22.

4. Ibid., p. 311.

5. OTA Report, pp. 349–51.

6. U.S. General Accounting Office, *Hazardous Waste: New Approach Needed to Manage the Resource Conservation and Recovery Act* (Washington, D.C.: GAO/RCED-88-115, July 1988).

7. Other studies of annual hazardous waste production in the United States have been carried out by the Office of Technology Assessment, the Environmental Protection Agency, and the Chemical Manufacturers' Association. Production estimates are fairly similar, ranging from 247 million metric tons (CMA) to 266 metric tons (CBO).

8. Benjamin A. Goldman, *Hazardous Waste Management: Reducing the Risk* (Washington, D.C.: Island Press, 1986), p. 41.

9. Samuel S. Epstein, Lester O. Brown, and Carl Pope, *Hazardous Waste in America* (San Francisco: Sierra Club Books, 1982), pp. 7–13.

10. U.S. Environmental Protection Agency, *Toxics in the Community: National and Local Perspectives* (Washington, D.C.: Government Printing Office, September 1990), pp. 56–57.

11. Congressional Budget Office, *Federal Liabilities Under Hazardous Waste Laws* (Washington, D.C.: Government Printing Office, May 1990), p. xix.

12. Ibid., p. xvi.

13. EPA, *Toxics in the Community*, Chapter 5.

14. Ibid.

15. Environmental Protection Agency, *Surface Impoundment National Report* (Washington, D.C.: Government Printing Office, 1983).

16. Congressional Budget Office, *Hazardous Waste Management: Recent Changes and Policy Alternatives* (Washington, D.C.: Government Printing Office, 1985), p. 21 (hereafter referred to as the CBO Report).

17. OTA Report, pp. 114–24.

18. The General Accounting Office suggested that industry noncompliance with RCRA reporting requirements could be attributed in large part to expense and technical complexity. This is discussed in a report titled *Hazardous Waste: Groundwater Conditions at Many Land Disposal Facilities Remain Uncertain* (Washington, D.C.: GAO/RCED-88-29, February 1988).

19. For further discussion of this point, consult *State of the Environment: A View toward the Nineties* (Washington, D.C.: Conservation Foundation, 1987).

20. Walter A. Rosenbaum, *Environmental Politics and Policy,* 2nd ed. (Washington, D.C.: CQ Press, 1991), p. 206.

21. See, e.g., "Congress Tightens Hazardous Waste Controls," *CQ Almanac 1984* (Washington, D.C.: Congressional Quarterly, Inc., 1985), p. 305.

22. See, e.g., Ann Bowman, "Superfund Implementation: Five Years and How Many Cleanups?" In Charles Davis and James Lester, eds., *Dimensions of Hazardous Waste Politics and Policy* (Westport, Conn.: Greenwood Press, 1988), pp. 132–33.

23. Bruce Williams and Albert Matheny, "Hazardous Waste Policy in Florida: Is Regulation Possible?" In James Lester and Ann Bowman, eds., *The Politics of Hazardous Waste Management* (Durham, N.C.: Duke University Press, 1983), pp. 92–93. See also U.S. Office of Technology Assessment, Superfund Strategy (Washington, D.C.: Government Printing Office, 1985).

24. Malcolm Getz and Benjamin Walter, "Environmental Policy and Competitive Structure: Implications of the Hazardous Waste Management Program," *Policy Studies Journal* (Winter 1980).

25. U.S. General Accounting Office, *Inspection, Enforcement, and Permitting Activities at New Jersey and Tennessee Hazardous Waste Facilities* (Washington, D.C.: GAO/RCED-84-7, June 1984).

26. See, e.g., Mark Henkels, "Duty and Discretion in a Wayward Agency: The U.S.

Environmental Protection Agency Implementation of the Hazardous Waste Laws," *Social Science Journal*, *25*, No. 1 (1988), pp. 53–65.

27. Steven Cohen, "Defusing the Toxic Time Bomb: Federal Hazardous Waste Programs." In Norman Vig and Michael Kraft, eds., *Environmental Policy in the 1980s* (Washington, D.C.: CQ Press, 1984).

28. Paul Dickson, "Citizen Risk Perception and Participation Strategies: Siting Centralized Hazardous Waste Management Facilities in the Northeast." Paper delivered at the 1983 Annual Meeting of the American Society for Public Administration, New York, New York.

29. David J. Webber, "Is Nuclear Power Just Another Environmental Issue?" *Environment and Behavior*, *14* (January 1982), pp. 72–83.

30. Richard N.L. Andrews, "Deregulation: The Failure at EPA." In Norman Vig and Michael Kraft, eds., *Environmental Policy in the 1980s* (Washington, D.C.: CQ Press, 1984), p. 163.

2

FORMULATING FEDERAL HAZARDOUS WASTE POLICIES

The formation of federal hazardous waste policies consists of actions designed to get the attention of elected officials, the process of gaining congressional approval, and the subsequent alteration or refinement of policy goals through legislative reauthorization. We begin by considering how a given issue is propelled from a position of relative obscurity to the decision-making agenda of Congress. Questions to be addressed in this section include the identification of key individuals and organizations responsible for providing the energy and resources to shape policy proposals as well as the strategies used to expand the organizational bases of program support. By analyzing the origins of the Resource Conservation and Recovery Act of 1976 (RCRA) and the Comprehensive Environmental Response, Compensation, and Liability Act of 1980 (Superfund), we can illustrate differences in agenda setting that are attributable to a combination of issue definition and historical circumstances.

After members of Congress have been persuaded that a proposal is indeed a legitimate topic of inquiry, efforts are to be made to develop and adopt a statutory package. Our concern in this section is with the strategic interplay between legislators, interest groups, and executive branch actors in marshalling support for the enactment of RCRA and

Superfund. Tactics employed in pursuit of these statutory goals have tended to vary with the alignment of legislators and groups opposed to the bills under consideration, the degree of public concern about the issue, and the presence or absence of competing program objectives.

We conclude the chapter with an analysis of legislative reauthorization. Many laws enacted by Congress (including RCRA and Superfund) are not permanent: They are adopted for a specific period of time and must be renewed by lawmakers. Shifts in program emphasis may reveal something about the institutional commitment of Congress to the policy objectives of RCRA and Superfund. Has change occurred in order to placate regulated interests that were unable to prevent the adoption of laws that in their view are unduly restrictive and costly? Or are the newer policies more responsive to the preferences of program advocates who seek to build on the original mandate by closing old loopholes and adding new responsibilities? Our examination of the Hazardous and Solid Waste Amendments of 1984 and the Superfund Amendments and Reauthorization Act of 1986 reveals elements of each, but the weight of the evidence tends to indicate that the environmental protection emphasis was strengthened.

SETTING THE AGENDA

While public officials are continually beset with demands for governmental action on a wide array of trivial and nontrivial problems, there is little likelihood that any of these problems will achieve agenda status until it has been converted into an issue. According to Eyestone, "An issue arises when a public with a problem seeks or demands governmental action and there is public disagreement over the best solution to the problem."[1] Once an issue is so raised, it may become part of a systemic agenda or an institutional agenda. The *systemic agenda* consists of issues that are perceived to be deserving of public attention by members of the political community and are consequently viewed as concerns within the proper scope of governmental authority.[2] The *institutional* or *governmental agenda* refers to proposals that are under serious consideration by authoritative decision makers in the public sector.[3]

Agenda setting may reflect a process in which issues are defined, converted to proposals, and ultimately expanded from the systemic agenda to the institutional agenda. The bottom-up or grass-roots approach implied by this sequence of events tends to place greater emphasis on the role of nongovernmental actors such as the media, the manipulation of symbols to enlarge organizational bases of support, and characteristics of the issue itself.[4]

An alternative approach offered by Kingdon is based on the observation that policy proposals are more apt to originate from a more visible cluster of participants, consisting of the president and his high-level appointees, prominent members of Congress, the media, and, on occasion, party leaders.[5] Because of the positional advantages wielded by legislative and executive officials, access to the institutional agenda is less encumbered—particularly if the president puts forward a set of policy priorities. Otherwise, the formal consideration of a given proposal is dependent upon the coupling of problems, policies, and politics; i.e., "a problem is recognized, a solution is available, the political climate makes the time right for change, and the constraints do not prohibit action."[6]

To what extent does the consideration of federal hazardous waste policies appear to fit either or both of these perspectives? Let us begin by examining the origins of the RCRA program for the regulation of toxic chemicals. Early consideration of government response to the control of hazardous pollutants occurred within a political climate that was quite supportive of environmental concerns. Public opinion was strongly on the side of governmental action to correct pollution problems. A nationwide Harris Survey conducted in 1971 found that 83 percent of the American public wanted the federal government to spend more money on air and water pollution-control programs.[7]

Jones has suggested that the policymaking conditions were so ripe for governmental action that members of Congress became less comfortable with their traditional majority-building approach based on the gradual accumulation of evidence to marshal support for a given policy proposal. Instead, between 1970 and 1972, legislators found it necessary to fashion policy responses to meet public demands for environmental legislation.[8] Moreover, the prospective adoption of these policies was not jeopardized by countervailing policy goals such as economic growth or energy independence, issues that did not become salient until the mid-1970s.

In addition, there was relatively little conflict between political parties or between the executive and legislative branches over environmental policy issues. Both President Nixon and Senator Edmund Muskie, the Democratic front-runner for the 1972 presidential nomination, actively courted the support of the environmental community.[9] Tough regulatory laws such as the Clean Air Act of 1970 and the Water Pollution Control Act of 1972 were adopted by overwhelming margins in Congress, receiving the backing of Republicans and Democrats alike. The creation of the Environmental Protection Agency (EPA) in 1970 through executive reorganization provided yet another institutional vehicle for the consideration of antipollution legislation.

Hazardous waste problems were initially analyzed within the context of solid waste policy. While the prospective regulation of toxic chemical byproducts was viewed favorably in some circles as a laudable attempt to tighten federal control over land- as well as air- or water-based pollutants, others argued that the appropriate locus of policy concern was state or local government. Congress eventually began work on the development of hazardous waste policy in the same fashion as the air and water pollution programs. Funds were initially authorized under the Resource Recovery Act of 1970 to study the problem.

Section 212 of this statute called for EPA to prepare a study detailing the means by which a system of national repositories for the storage and disposal of hazardous wastes could be created and to submit the report for congressional consideration by October 1972. The term "hazardous waste" was not explicitly defined in the statute but was said to embrace "radioactive, toxic chemical, biological and other wastes which may endanger public health or welfare."[10]

The report was forwarded to Congress by the EPA in June 1973. In general, the document indicated that government had done an inadequate job of managing hazardous residues and that federal and state legislation to compel waste generators to improve their management practices was "spotty or nonexistent."[11] Thus, producers of hazardous wastes were under little or no pressure to commit financial resources to control toxic byproducts more responsibly; indeed, to do so would place the company at a competitive disadvantage economically. A policy that focused only on the siting of federal treatment, storage, and disposal facilities would do little to address the underlying problem of inadequate management practices. The authors of the EPA report recommended that a national hazardous waste regulatory program be established to ensure "environmental protection, equitable cost distribution among generators and recovery of waste materials."[12]

This document provided sufficient information and analysis to attract the attention of key legislators in the 93rd Congress. While selected members of Congress and committee staffers supported the idea of a national hazardous waste management program, EPA officials were equally responsible for taking the initiative in shifting the analytic focus from one of distributive policymaking (i.e., providing low-cost storage capacity for industry disposal of hazardous waste on public lands) to one of regulation.[13] There was little, if any, promotion of this proposal by the media or by environmental organizations, whose energies were directed at more visible policy issues—namely, the Toxic Substances Control Act.[14]

Consideration of RCRA thus coincides with Kingdon's characterization of agenda setting. The proposal was put together by "hidden specialists" within EPA and congressional committees with expertise in environmental policy rather than nongovernmental political actors.[15] The content of the EPA document reflects another dimension of policy agendas—the importance of recombination (repackaging older ideas into a new format) in relation to mutation (coming up with ideas that are original and unique).[16] The hazardous waste regulatory framework recommended by EPA combined the "cradle-to-grave" tracking system originally developed in Western Europe with some features of the goals and timetables enforcement system used in the air and water pollution-control laws.[17]

Getting Superfund onto the public agenda proved to be more politically complicated than was true of RCRA. Shortly after the adoption of RCRA (to be discussed in the following section), information about abandoned dump sites began to trickle in and EPA officials realized that they had underestimated the magnitude of the policy problem. A report prepared by a consulting firm under contract to EPA concluded that 32,000 to 50,000 hazardous waste disposal sites were scattered throughout the United States and that 1,200 to 2,000 of these might be classified as "extremely dangerous."[18] The report also indicated that 500 to 800 of the sites either were abandoned or remained under the control of owners who could not afford to clean them up.[19] Moreover, these problems could not be resolved through the implementation of existing laws.[20] Additional statutory authority was clearly needed.

Despite mounting evidence that chemical contamination of land and water was more serious and widespread than previously acknowledged, agenda status for a Superfund bill was by no means assured. Public support for environmental policy initiatives remained fairly high; however, the emergence of competing policy issues suggested that the kind of unswerving bipartisan approval characteristic of the 1969 to 1972 period was unlikely to occur. In his analysis of public opinion toward environmental issues, Dunlap concluded that a modest decline in support in the latter half of the 1970s could be attributed to "continuing concerns about energy supplies throughout the decade, a worsening economic situation and a taxpayer's revolt begun by California's Proposition 13 in 1978."[21] Public officials became increasingly aware of the costs associated with the imposition of pollution-control regulations and realized that appeals based on the desire for clean air or water alone would no longer suffice as a call to action.

However, it is also unlikely that support for environmental programs can be significantly diluted over time by the desire for economic growth or increased energy supplies. This is particularly true if an issue becomes timely because of "triggering devices" such as a natural disaster or technologically induced changes in the environment.[22] In 1978, environmental and public health problems posed by the improper disposal of hazardous wastes were demonstrated in a rather dramatic fashion through media coverage of several pollution spills.

The most widely publicized incident took place in Love Canal, a neighborhood within Niagara Falls, New York. Chemicals began to leach from the ground and into basements near a school in a residential area that formerly had been used as a hazardous waste dump site by Hooker Chemical Company. Residents increasingly obtained information—much of which was anecdotal—suggesting that the incidence of cancer and birth defects was unusually high in the areas with the highest concentration of pollutants.

The rise of public concern was precipitated by investigative journalism in the early stages of the crisis. However, Love Canal's tenure as a news item and subsequent expansion to the national media was attributable to the evolving relationship between the press and an ad hoc neighborhood association headed by Lois Gibbs.[23] Association members spent considerable time learning about hazardous wastes and their effects on human health; the interpretation of scientific data and the assumptions underlying their use; and, most importantly, ways of getting and maintaining access to the media.[24] The message Gibbs and her followers sought to convey was that governmental action and resources were urgently needed to ameliorate risks to human health.

In addition to the Love Canal tragedy, a number of other environmental horror stories were uncovered; two incidents were especially newsworthy. Approximately 17,000 barrels of chemical wastes were discovered at a site near Louisville, Kentucky (later dubbed Valley of the Drums), and 54 public wells providing drinking water for 100,000 people in Long Island, New York, were found to be contaminated with pollutants.[25] Soon EPA was inundated with calls from members of Congress wanting to know what the agency intended to do about these incidents.

EPA Administrator Douglas Costle might have responded with a series of small-scale emergency actions permitted under the statutory authority of RCRA. Instead, he chose to "maximize public perception of the dangers associated with abandoned hazardous waste sites and to exploit that awareness in order to win approval of a complex and expensive new regulatory program."[26] Regional EPA officials were told to prepare a list of the ten worst dump sites within their jurisdictions and to make these lists available to the media.

The combined effects of public concern over abandoned dump sites and media coverage resulted in the development of Superfund policy proposals by the Carter Administration and key members of Congress in 1979. A comparison of Superfund with RCRA suggests that in each case EPA emerged as a key player in fashioning policy alternatives. On the other hand, Superfund followed a different route to the governmental agenda in several respects. Clearly, the policy problem of abandoned dump sites was much more visible to Congress, affected industries and the general public than was the little-noticed regulatory program conceived within the RCRA umbrella. Nongovernmental actors such as the media and environmental organizations were more active in publicizing events, that in turn produced a snowball effect which was difficult for elected officials to ignore.

EPA also had a hand in expanding the issues by placing greater emphasis on the public health risks than on environmental degradation. Accelerated activity on behalf of Superfund was important to staffers anxious to ride the wave of public concern to maximum political advantage. As long as accidental pollution spills remained fresh in the minds of the public and elected officials, the political pressure to enact hazardous waste legislation was likely to eclipse economic arguments centering on regulatory compliance costs.

Industries targeted by the proposed legislation were equally determined to defeat or at least water down the bill. A link between rising regulatory compliance costs and worsening economic conditions was repeatedly made by industry representatives as well as presidential candidate Ronald Reagan. In short, the attainment of agenda status resembles a mixed model of sorts, in which insiders within EPA and Congress pushed for legislative consideration of Superfund with ample reinforcement from grass-roots activists and the press.

POLICY ADOPTION

The adoption of public policies is preceded by the development and consideration of proposals. An especially important concern is whether there are discernible political biases embedded within congressional committees with jurisdiction over hazardous waste issues. Are committee members, on balance, inclined to support the use of federal authority to enhance environmental protection? Or is the desire to provide statutory coverage inhibited by the belief that other policy objectives (such as the development of energy resources) are more pressing or the position that environmental goals are best achieved by state rather than federal officials? Also, does the pattern of decision reflect a decidedly partisan or regional orientation?

Some committees are clearly preferable to environmental program advocates because of key individuals who can be counted on for effective and sustained effort on behalf of selected policy proposals. For example, Senator Edmund Muskie (D–Maine) became well known as a champion for ecological policy issues in the 1960s and 1970s, while Representative Henry Waxman (D–California) has been quite active from the 1980s to the present.[27]

Building support for a policy proposal also depends on the manner in which issues are defined and molded into policy proposals. If access to the governmental agenda is achieved through the grass-roots model, greater emphasis is placed on using popular symbols (such as pictures of the fur seal) to attract the broadest possible coalition of supportive interest groups.[28] According to Anderson, pollution control is an attractive issue, since:

> It affects everyone....Moreover, it is difficult to oppose pollution control because one cannot win many political allies by openly favoring dirty air and water. In addition, pollution control is often tied to public health, which is another popular concern. As the old song has it, "Everybody wants to go to Heaven but nobody wants to die."[29]

In short, an effort is made to win the vote of a legislator for a given program by emphasizing its relationship to constituency interests and to public opinion writ large. On the other hand, policy proposals generated within government may result in the assignment of greater weight to other decision-making criteria, such as deference to the expertise of colleagues within the legislature, personal values, or, in some cases, political party affiliation.[30]

Resource Conservation and Recovery Act

Congressional deliberation over the hazardous waste regulatory policy began in 1974 with hearings before the House Committee on Interstate and Foreign Commerce and the Senate Committee on Public Works. Perhaps the most active legislator on this issue was Representative Paul Rogers (D–Florida). As chairman of the Subcommittee on Public Health and the Environment, he favored a major role for EPA in the resolution of solid waste management problems.

In his opening statement before the subcommittee, Rogers was critical of the failure of Nixon Administration officials to provide sufficient funds and staff to EPA for implementing the Resource Recovery Act of 1970, a solid waste program.[31] He emphasized the serious nature of waste management problems, particularly the increase in the number of open dumps in the United States. Yet another example used to illustrate the low priority of this issue was the *1973 Annual Report of the Council of Environmental Quality*, which devoted only 11 of its 1,200 pages to this topic.[32] In a cumulative sense, these events suggested a rather unfortunate turn of events to Rogers, who concluded that "the EPA, perhaps because of the action taken by OMB, seems to be expending more effort in disbanding the [Resource Recovery Act] than it is in administering it."[33]

The Nixon Administration exhibited an ambivalent stance toward the newly proposed Resource Conservation and Recovery Act. Handling solid wastes was considered to be an overly extravagant use of increasingly scarce federal funds. To White House officials, this exemplified the type of policy that could not and should not be administered from Washington. Accordingly, substate or local governments were viewed as the appropriate loci of both policy control and operating expenditures.[34] On the other hand, administrators within the White House and the OMB were inclined to support the hazardous waste management program as a separate package.

Similar hearings were held by the Environmental Pollution Subcommittee of the Senate Public Works Committee, chaired by Senator Jennings Randolph (D–West Virginia). Once again, policy differences between the Nixon Administration and members of Congress were found. However, no further action was taken on this bill in the 93rd session, since both the House and Senate committees with jurisdiction over solid waste policy issues were preoccupied with the oil crisis and with proposals to ease environmental regulatory constraints on domestic energy production (such as the relaxation of clean air standards).[35]

By 1975, the climate for policy change was altered by the departure of President Nixon. His pardon by President Ford in the wake of the Watergate scandal had a pronounced effect on the midterm elections of 1974. The Democrats were able to obtain sizable majorities in both houses of Congress. The actual content of solid and hazardous waste policy was subsequently influenced by two developments. First, Congress was able to craft a statute independently of executive branch involvement. Because of his opposition to the generation of any new domestic programs, President Ford refused to resubmit the limited policy package authored by EPA Administrator William Ruckelshaus in the preceding session. As a result, legislators felt less obligated to seek input from administration officials.

Second, Congress was simultaneously considering a similar pollution-control policy proposal, which served to divert the attention of industry and environmental groups. The Toxic Substances Control Act (TSCA) called for premarket testing of new products and was clearly a priority issue for chemical firms represented by the Chemical Manufacturers' Association (CMA). CMA strategists recognized that some form of legislation was inevitable and thus undertook a major initiative to minimize governmental impacts on industrial operations and trade secrets. The degree of concern expressed by industry or environmental organizations in the hazardous waste sections of the solid waste bill was comparatively low.[36]

The window of opportunity for the congressional adoption of RCRA finally opened in mid-1976, when the House and the Senate approved related bills dealing with hazardous waste controls. The bills differed in three ways—the Senate version authorized more money and provided funding for recycling and recovery projects, while the House of Representatives' package called for an expansion of state and regional solid waste research and development programs. Since the date for adjournment was drawing near, a compromise bill was worked out by House and Senate committee staffers. They split the difference on program funding, dropped the Senate provision on loan guarantees for recycling and recovery programs, and incorporated the House section dealing with solid waste research and development.[37] As amended, RCRA was adopted with little fanfare by both houses of Congress and was signed into law by President Ford on October 22, 1976.

Subtitle C of RCRA addressed the hazardous waste management problem by establishing an elaborate "cradle-to-grave" regulatory framework.[38] Hazardous wastes were defined in both generic terms (as ignitable, corrosive, reactive, or toxic) and by listing specific wastes and industrial waste streams. A tracking system also was incorporated to monitor the whereabouts of toxic chemical substances from the production phase to their ultimate treatment, recovery or disposal. Any company involved in the generation or transport of these wastes is required to prepare a manifest for record keeping and reporting purposes. Firms handling hazardous wastes at any stage of the process are also required to obtain a permit from the EPA or an authorized state agency.

Superfund

Congressional action on Superfund was prompted by media coverage of the Love Canal incident and related events throughout the United States. In March 1979, hearings were held in both houses of Congress. At the same time, considerable posturing was taking place among political actors in industry, environmental organizations, and the media. Newspaper editorials chastised chemical firms for irresponsible waste management practices and called for a fuller disclosure of existing problems as well as a willingness to participate financially in the cleanup of abandoned dump sites.

Industrial spokespeople responded by arguing that the costs of ameliorating risks to public health posed by these sites should not be borne by the corporate sector alone. Because consumers benefited from lifestyle improvements made possible through chemical technologies, they should be expected to foot part of the bill. According to Robert Roland, head of the Chemical Manufacturers' Association (CMA):

> The solid waste problem, including toxic or hazardous waste, is not just the problem of the chemical industry. It is a result of society's advanced technology and pursuit of an increasingly complex lifestyle....Everyone

> should realize that the blame does not belong to a single company or a single industry but to all of us as individuals and as an advanced society. Rather than looking for scapegoats, we should recognize the dilemma and consider new ways to encourage the disclosure of dump site information and ways to limit the crushing liabilities that could result.[39]

Finding ways to work hand in hand to resolve hazardous waste management problems did not, in his view, include a public policy solution. CMA received considerable pressure internally from major chemical firms to adopt a hard line against any Superfund proposal, especially a bill containing strict liability provisions or a funding formula largely based on industry contributions. Companies opposing legislation of any sort included Dow, DuPont, Allied Chemical, and Union Carbide, while Monsanto, Rohm & Haas, and Olin successfully pushed for a more conciliatory stance that would allow industry greater say in shaping policy outcomes.[40]

President Carter was determined to get Superfund legislation enacted. One of his chief concerns was coming up with some type of solution for the compensation issue. Should Superfund be financed primarily from general tax revenues or from a more specific tax levied upon chemical industries? To address this problem, a conference was convened in the spring of 1979. Those attending included the president and representatives from EPA, the Council on Environmental Quality, and other federal agencies.

Agenda items under consideration included a fund created wholly from general governmental revenues and a fund based on financial contributions from both industry and government. Individuals representing the Department of Commerce, the Council on Wage and Price Stability, the Office of Management and Budget, and the Council of Economic Advisors pushed for public funding, while a shared approach was favored by administrators from EPA, the Council on Environmental Quality, the Department of Health and Human Services, and the Occupational Safety and Health Administration. In the final analysis, President Carter spoke out in favor of a funding formula based on an 80 percent industry/20 percent government split. The "polluter pays" principle thus became a cornerstone of the Carter Administration's Superfund proposal.[41]

The political and strategic rationale for an industry-based tax was developed by EPA Administrator Douglas Costle and Thomas Jorling, Assistant Administrator for Water and Hazardous Waste. Each recognized the long-term difficulty of relying on Congress to provide a consistent and adequate amount of funding for a Superfund program because of the "issue attention cycle."[42] The cleanup of abandoned dump sites was likely to be both time-consuming and expensive, and neither Costle nor Jorling wished to risk placing ongoing decontamination efforts on hold because of uncertainty in Congress over annual program funding decisions. They wanted a funding mechanism that not only would be less susceptible to change over time but also would be politically insulated from competition with other policy issues for increasingly scarce federal monies.

A related question was whether some industries would feel the financial bite more than others. EPA proposed that the tax be placed on firms responsible for the production of chemical feedstocks used in the development of compounds with hazardous byproducts. Such an approach contained several administrative advantages:

> The feedstock approach would involve fewer than one thousand collection points. More importantly...feedstocks were produced by large stable chemical companies, not by small marginal firms. Therefore, this approach provided a far more stable source of revenue. Finally, since the revenue source was divorced from the actual generation and disposal of hazardous waste, it provided no incentives for firms to opt out of the RCRA control program.[43]

Other substantive areas of concern considered by EPA officials and members of Congress—much to the dismay of CMA—were liability and victim compensation issues. Senators Muskie and Culver (D–Iowa) were aware that existing law (as of 1980) made it extremely difficult for victims or environmental regulatory agencies to recover damage claims from firms thought to be responsible for the improper storage or disposal of hazardous wastes. The epidemiological standard of evidence in U.S. courts clearly worked against the interests of victims, since they had to prove that illness or other medical problems could be directly linked to exposure to a given toxic chemical rather than lifestyle or workplace risks.[44]

What, then, did policymakers offer as policy options? One idea under discussion was diverting a portion of Superfund monies to compensate victims of toxic chemical exposure for medical and economic costs incurred as well as dump site cleanup costs. Another proposal extended "joint and several" liability to toxic chemical substances, thereby ensuring that firms contributing to the contamination of land on or near an abandoned dump site could not evade financial responsibilities associated with cleanup costs simply because other firms had dumped the same chemical byproducts at the same location.[45]

Superfund supporters were unable to push legislation through Congress in 1979, but they renewed their efforts in the spring of 1980. In the House of Representatives, a bill under consideration before the Transportation Subcommittee of the Commerce Committee received a political boost from a tragic incident. On April 22, a warehouse on the premises of Chemical Control Corporation in Elizabeth, New Jersey, caught fire and exploded, killing several workers. One week later, Representative James Florio (D–New Jersey) persuaded members of his subcommittee to approve a bill calling for a $600 million fund, based on a 50-50 split between industry and government, plus the provision of strict liability standards. The bill was adopted by the Commerce Committee two weeks later.

Industry officials pushed to get Representative Florio's bill referred to the Ways and Means Committee, ostensibly because of its revenue provisions. In addition, industry strategists believed that their concerns would receive a more sympathetic hearing.[46] However, after the proposal was actually sent to this committee, the membership reacted in a less predictable fashion. Much to the surprise of industry leaders, legislators were persuaded to enlarge rather than reduce the role played by industry in the cleanup of abandoned hazardous waste dump sites. Representative Thomas Downey (D–New York) argued successfully that the size of the fund should be increased to $1.2 billion and that industry's relative contribution to the fund should be raised to 75 percent. One analyst suggests that the action taken by the Ways and Means Committee was primarily "an effort to protect the Treasury Department and ensure that industry would carry an adequate share of the burden to spare government the costs of cleanup."[47]

House advocates of Superfund benefited from continuing media attention to the issue and from a strategic error made by CMA President Robert Rowland. On September 11, Rowland appeared on ABC's *Nightline* and indicated that the CMA supported the Florio bill. This statement confused the hard liners within the association, who had been led to believe that the amendments made by the Ways and Means Committee were overly punitive, unduly expensive, and therefore unacceptable. The following day, Rowland issued a retraction, stating in a letter that his intent had been to convey CMA's support for the $600-million fund contained within the earlier version of the bill adopted by the Transportation Subcommittee. However, a loss in credibility was an unintended effect of his "misstatement."[48]

Superfund supporters within the House of Representatives successfully resisted a pair of potentially crippling amendments. One would have deleted industry contributions to the fund, while the other would have required greater congressional oversight of EPA decisions. Both were easily defeated in voice votes and on September 27, the House of Representatives overwhelmingly passed the Florio bill by a large margin.

Early work on a Superfund bill by Senators Culver and Muskie demonstrated a willingness to tackle some of the tougher liability issues. Their proposal gave victims of toxic substance exposure the right to obtain compensation for injuries or economic losses. It also called for a shift in the burden of proof from government to industry. Under their plan, legal action could be initiated by individuals able to show a "reasonable relationship" between exposure and economic or medical damages.

Proponents of this bill received a setback in early 1980, when President Carter appointed Senator Muskie to be his new Secretary of State. Because of his stature as the chief architect of federal pollution-control laws dating back to the mid-1960s, Senator Muskie's views on environmental policy matters carried considerable weight with fellow legislators. However, the leadership role was picked up by Senator Robert Stafford (R–Vermont), who worked with Senator Culver to reshape the bill into a more politically acceptable document.

How was this accomplished? Some toxic substances were exempted from statutory coverage, but the size of the fund ($4.1 billion) was substantially greater than that found in competing legislative proposals. In addition, President Carter sent a letter of support for the bill to Senator Jennings Randolph, chairman of the Environment and Public Works Committee. The bill was reported out of committee on June 27.

The next legislative hurdle proved to be more formidable. The Senate Finance Committee gained jurisdiction to review the bill. Industry officials were optimistic that their views would receive favorable consideration, since the committee chairman, Senator Russell Long (D–Louisiana), had long championed the policy interests of oil and chemical companies. Several committee members were opposed to Superfund on ideological grounds, while others were concerned about the economic impact of prospective legislation on smaller companies.[49] The possibility of defeat through delay became a live prospect, since Congress was considering adjournment in early October.

An opportunity for movement on Superfund emerged when Congress chose to return after the presidential election for a "lame-duck" session. Prospects for adoption were uncertain at best, since a number of key supporters, notably President Carter and Senator Culver, had failed in their bids for re-election. Moreover, the incoming administration of President-Elect Ronald Reagan had strongly pushed a variety of campaign themes emphasizing deregulation; hence, most political observers considered it extremely unlikely that support for expensive new environmental policy initiatives would be forthcoming. In short, Superfund advocates within the Senate recognized that legislative approval clearly required a willingness to compromise.

Several last-minute maneuvers led to the eventual passage of Superfund by the Senate. First, Senator Long scheduled a vote within the Senate Finance Committee, and in mid-November they voted to send the bill forward without a formal recommendation. Second, Senator Stafford had anticipated the possibility of favorable action and had prepared a compromise proposal with several changes designed to accommodate industry concerns. The size of the fund was scaled back considerably, provisions for joint and several liability were deleted, and the section on victims' compensation was eliminated. These incentives were tempered by a threat to introduce a "more stringent" Superfund bill the following year if chemical firms maintained a hard-line political stance against any form of regulation by EPA.[50]

In addition, cracks in the chemical industry coalition began to appear. The chief executive officers of DuPont, the largest chemical firm in the United States, and Union Carbide, the third largest, expressed their support for some form of Superfund legislation, leaving CMA President Rowland in the uncomfortable position of continuing the fight on behalf of the chemical industry without the backing of some of its most influential spokespeople. A final logistical hurdle was overcome when Senator Jesse Helms (R–North Carolina) agreed to end his active opposition to the bill in exchange for a further reduction in the size of the fund, from $2.7 billion to $1.6 billion.

These changes ensured a favorable legislative outcome. As amended, the bill was adopted by voice vote in the Senate. The House of Representatives then agreed to accept the Senate version of the bill, effectively deleting a section that would have provided statutory coverage for oil spills. President Carter formally signed the Comprehensive Environmental Response, Compensation, and Liability Act on December 11, only weeks before Ronald Reagan took office.

In brief, Superfund provided a $1.6-billion fund to finance hazardous waste cleanup actions at abandoned dump sites over a five-year period. Monies were derived from a tax on chemical feedstock producers (87.5 percent) and, to a lesser extent, from federal revenues (12.5 percent). EPA was assigned the responsibility for preparing a National Priority List (NPL) of the most dangerous (from a public-health perspective) hazardous waste dump sites throughout the United States, including at least one site from each state. Funding could be provided for activities such as the short-term or permanent cleanup of hazardous waste dump sites, epidemiological studies, chromosomal screening, and a registry of persons exposed to toxic substances.

STATUTORY REAUTHORIZATION

The reauthorization of public policies reflects a peculiar combination of institutionalization and change. On the one hand, reaffirming existing policies is often easier than marshalling the necessary support to initiate new policies. Organization and staff are in place, fiscal resources have been allocated, and constituencies designed to receive program benefits and to provide ongoing political support have been developed and presumably nurtured. Continuity in the design and implementation of programs is achieved and subsequent shifts in policy are typically incremental and short-term.[51]

On the other hand, some policies are characterized by a sizable expansion in scope and resource allocation. Change is fundamental rather than incremental. It is often accompanied by the acquisition (since the passage of the initial policy) of new policy-relevant information indicating that the problem is more serious than the policymakers realized.[52] Of equal or greater importance is the saliency of the policy problem to the public and policy actors in relation to other policy issues.[53] Action on the reauthorization of governmental programs is spurred by events such as an environmental catastrophe or by an increase in public concern about the ability or willingness of public officials to enforce existing policies.

An examination of congressional efforts to reauthorize RCRA and Superfund in the early and mid-1980s reveals the emergence of several issues that represented a marked departure from the original statutes. Some of the more pressing concerns included the degree of risk associated with exposure to toxic substances, the number and distribution of contaminated dump sites, costs related to program use and the prospective adoption of alternative policy or management options, the need to construct a list of hazardous wastes recognized by policymakers at all levels of government, and the identification and reduction of administrative overlap between related pollution-control policies, among others.

The number of issues added to program reauthorization decisions was large and could be at least partially attributed to the nature of the policy problem. Legislators tended to perceive hazardous waste as a problem that combined public-health risk factors with a high degree of technological uncertainty.[54]

Reauthorizing RCRA

Following the adoption of RCRA in 1976, it soon became evident that EPA was experiencing major difficulties obtaining necessary information on the number of disposal sites, the relationship between improper waste disposal practices and groundwater quality, and alternative approaches to the disposal of toxic wastes.[55] Such problems contributed to EPA's failure to meet the 1978 deadline for the promulgation of interim regulations. In addition, members of Congress became increasingly critical of EPA for devoting too few resources to hazardous waste programs and for its inability to appease constituents worried about the identification and control of dump sites.[56] Despite these reservations about EPA's performance, RCRA was reauthorized in 1980 with a few minor changes. Agency enforcement powers were strengthened somewhat and, in a compensatory gesture to industry, Congress decided to exempt from regulatory coverage certain types of waste (such as drilling muds used in oilfield operations).

In 1982 and 1983, Congress sought and obtained information on the dimensions of hazardous waste as a policy problem and the kinds of options that would enable EPA and state-approved regulatory agencies to better manage the recovery, storage, or disposal of such wastes. The Office of Technology Assessment (OTA) released a major study in 1983 that was quite critical of existing management practices at EPA. The OTA report recommended that policymakers address several problems, including inconsistencies between state and federal programs, inadequate and inconsistent data on the amount and type of waste generated annually, the scope and distribution of pollutants produced by small-quantity generators (exempted from coverage under the 1976 RCRA law), and the lack of producer incentives to consider pretreatment or recycling of wastes in lieu of land-based containment.[57]

A related study released by the National Academy of Sciences (NAS) in 1983 focused on waste disposal and recovery options. Emphasis was placed on the desirability of on-site source reduction by chemical firms and treatment technologies designed to reduce the toxicity of hazardous chemical wastes.[58] The combined effect of the NAS and OTA studies was to reveal a fundamental problem which Congress and EPA had failed to address—the ability of waste-generating industries to avoid the true costs of managing pollutants. Or, as one analyst has indicated, "Cleanup costs were deferred since there were no marketplace incentives to invest in the technology required to effectively recover, recycle or dispose of hazardous wastes."[59]

Movement in Congress to significantly revamp RCRA was accelerated not only by the information contained in these reports but by a series of events that placed hazardous waste management problems squarely in the national limelight. An environmental dispute in Times Beach, Missouri, attracted media attention in late 1982, following warnings from Dr. Barry Commoner and the federal Centers for Disease Control that community residents should not be exposed to dioxin for long periods of time. In 1972, dirt roadways had been sprayed with waste oil containing a large amount of dioxin as a means of controlling dust, and subsequent analyses conducted in 1982 indicated that a sizable concentration of the substance remained in the soil.[60] EPA responded by ordering the evacuation of residents, and it later purchased dioxin-contaminated homes within the affected area.

Attention to hazardous waste issues was also affected by the outbreak of scandal in early 1983 over the political manipulation of hazardous waste program decisions at EPA. Major charges included the use of political criteria to guide the release of Superfund cleanup monies; conflict of interest and industry favoritism claims lodged against Rita Lavelle (an assistant administrator responsible for the management of hazardous waste programs); and the use of a "hit list" by agency officials to identify environmental extremists with EPA who were subsequently targeted for transfer, demotion, or dismissal.[61] No fewer than six congressional committees conducted investigations of EPA activities and a consequence of the controversy was substantial turnover of top-level agency personnel—13 officials, including Lavelle and EPA Administrator Anne Burford, resigned.

Within the context of media attention to hazardous waste issues and the availability of policy-relevant information, Congress began deliberating on the reauthorization of RCRA. The House Committee on Energy and Commerce adopted a bill in May 1983 that contained several major changes. Two of the more notable features included the extension of regulatory coverage to underground storage tanks containing hazardous chemical substances or petroleum products and to small-quantity generators (companies generating between 100 and 1,000 kilograms of hazardous waste on a monthly basis). Another key provision called for an emphasis on recycling and forms of waste disposal other than land-based containment.

Other points of departure reflected a rather ambivalent view on the part of committee members. The bill strengthened EPA's enforcement authority but also reflected congressional reaction to ongoing political events by restricting EPA's discretion. This was particularly evident in the House adoption of a so-called "hammer clause," which imposed a ban on the land disposal of solvents and dioxin unless EPA was able to demonstrate ways of storing these substances safely within a 24-month time period.[62] On balance, these changes were favored by environmental organizations, while industry interests viewed the bill as an overly intrusive approach to environmental management. The House of Representatives adopted the bill in November 1983.

Senate action on RCRA took place within the confines of the Environment and Public Works Committee. A bill was reported out of committee in July 1983, but final Senate action on the bill was deferred for nearly a year. The chief hurdle to Senate approval was the federal OMB, which tempered its statement of support with objections to certain "inflexible and unnecessary regulatory mandates which could impose from $10 billion to $20 billion per year in added costs on the economy while providing limited environmental benefits."[63] Senate Majority Leader Howard Baker finally agreed to schedule floor debate on the bill after he had been presented with a petition signed by 52 senators from both parties requesting action. After acceding to a number of amendments granting concessions to members with specific concerns, the Senate passed the bill in mid-1984.

A conference committee consisting of senators and representatives who had worked on changes to RCRA convened in September 1984. They attempted to iron out differences between the two chambers and forge a compromise bill acceptable to all parties. Most of the bill was devoted to the closure of loopholes in the original legislation and was agreed to with little fanfare. Ten days of fairly intense negotiations ensued over three of the newer changes—bringing small quantity generators under the jurisdiction of RCRA, placing limitations on waste disposal in landfills, and regulating underground storage tanks. Interchamber differences were resolved and conferees agreed to drop a controversial plan to tack on a number of amendments pertaining to the Superfund program.[64] The bill easily cleared Congress and in November 1984, President Reagan signed the Hazardous and Solid Waste Amendments (HSWA) into law.

From a political perspective, the passage of HSWA reflects the salience of environmental policy concerns within an era of fiscal austerity. Environmental representatives were generally pleased with the substantive content of the law and there was bipartisan support for its adoption. Existing points of disagreement between the Reagan Administration (primarily the OMB) and Congress centered on the price tag associated with regulatory compliance; however, the political benefits of responding to hazardous waste policy problems clearly outweighed the costs. In his analysis of public attitudes toward the environment in the early 1980s, Mitchell argues that President Reagan essentially misinterpreted public opinion:

> In its belief that it has a mandate to de-emphasize environmental priorities in the name of economic recovery and less government, the Reagan Administration disregarded the burden of evidence taken from polls in the late 1970s; it fell into the "salience" trap of believing that the absence of widespread spontaneously expressed concern about the environmental issues signified a weakening in the environmental consensus....Following the storm of controversy over EPA's toxic waste program, public opinion polls in 1983 indicated that the President was not held personally responsible for the problems at EPA; but pluralities disapproved of his environmental programs and appointees and majorities expressed the belief that he shared his appointees' pro-business views.[65]

The politics of reauthorizing RCRA in 1984 and the subsequent reauthorization of Superfund in 1986 demonstrated that Congress as an institution was able and willing to seize the initiative in an area of policy largely ignored by the Reagan Administration.

An assessment of policy change from RCRA to HWSA illustrates the willingness of Congress to incorporate new information in statutory reauthorization decisions. Studies of RCRA implementation by the General Accounting Office and others found deficiencies in the original legislation that were subsequently addressed in 1984. First, Congress revealed a preference for industry to pursue methods of hazardous waste disposal other than land-based containment. Noncontainerized or bulk liquid hazardous waste was banned from disposal in landfills along with a list of wastes identified as hazardous within California's management program. EPA also was required to publish, within 24 months, a schedule for determining the appropriateness of a ban on the land disposal of other wastes listed under RCRA, beginning with substances posing the greatest threat to public health and the environment.

Second, Congress responded to reports directing greater attention to environmental problems created by smaller waste generators. The amount of waste exempted from regulatory coverage was reduced from 1,000 kilograms (2,200 pounds) to 100 kilograms (220 pounds) per month. In addition, EPA was empowered to regulate companies producing less than 100 kilograms of waste per month if it was deemed necessary to protect public health.

Third, owners of underground storage tanks containing petroleum or other hazardous substances were required to meet new regulatory guidelines. EPA was instructed to issue regulations requiring continuous monitoring by owners to ensure early detection of leaks, record keeping, reporting and correction of leaks, and performance standards for new tanks.[66]

Reauthorizing Superfund

Efforts to reauthorize Superfund began in earnest after President Reagan submitted a $5.3 billion proposal to Congress in February 1985. The emergence of subsequent disagreement over the size and direction of policy between the president and Congress was foreshadowed by differences in their respective assumptions concerning problem scope. EPA estimated that approximately 2,000 waste sites were dangerous and in need of cleanup action, while the congressional Office of Technology Assessment concluded that 10,000 sites represented a more accurate figure.[67] Interbranch disagreement over problem magnitude was eventually reflected in Superfund program preferences, which varied considerably in terms of policy content and cost.

Senator Stafford took the lead in shaping the Senate Superfund proposal, which was reported out of the Environment and Public Works Committee in March. It incorporated a number of items considered but rejected within the original law, such as a provision for victim compensation (an experimental program calling for the expenditure of $30 million over five years in five to ten geographical areas chosen by EPA), the inclusion of the Department of Defense and other federal agencies within the statutory umbrella, and an amendment allowing citizens or organizations to sue EPA for failing to carry out its programmatic responsibilities.

Other committees subsequently obtained jurisdiction to review and amend the bill. A five-year spending package of $7.5 billion was pieced together by Finance Committee members, who sought funds from diverse sources. New monies would be obtained from a new, broad-based business tax levied on the sale of manufactured and raw goods (excluding exports and unprocessed agricultural, food, and timber products). Other funds would be derived from three sources—a tax on chemical feedstocks, general tax revenues, and monies recovered from companies contributing to the pollution of hazardous waste sites.

Floor debate was quite contentious on several aspects of the bill, particularly the use of the broad-based business tax to provide much of the revenue for Superfund. The Reagan Administration strenuously objected to the tax and OMB officials responded by sending a letter to the Senate leadership threatening a presidential veto. A floor fight was narrowly averted when the Senate approved a "sense of the Senate" resolution authored by Senator Jesse Helms, which permitted the tax provisions to remain within the bill but called for a future House-Senate Conference "to report legislation containing a reliable financing mechanism for the Superfund program which does not contain the value-added tax."[68] Other amendments to the bill were also accepted and, in late September, the Senate approved the measure. Notable changes in the final bill included the deletion of the victims' compensation provision and the strengthening of a section giving communities a "right to know" about hazardous chemicals contained in nearby plants or facilities.

House action on a Superfund bill commenced within the Environment and Commerce Committee, which approved a bill in late July after several days of heated debate. All of the dissenters were Democrats, including Representative James Florio, one of the original architects of the measure. At issue was the mandatory schedule for the cleanup of abandoned dump sites on the NPL list. Since there was no effective deadline for the completion of cleanup studies, which would precede cleanup action, Representatives Florio and Henry Waxman contended that a timetable for cleanup actions was meaningless. A coalition of oil-state Democrats and Republicans successfully resisted an amendment to create a deadline for the completion of cleanup studies.

The bill gradually moved through a maze of committees prior to floor debate. A compromise bill was eventually adopted by a sizable margin in early December, after the resolution of disagreements over the business tax (deleted in lieu of higher levies to be imposed on waste-generating firms) and a "right-to-know" clause requiring industries to disclose information to workers about the discharge of chemicals known to be toxic or carcinogenic (adopted).

The House and Senate began meeting in conference in February 1986 to reconcile their differences. The chief hurdle was reaching an agreement on a funding approach for Superfund. Pressure to develop a workable solution was reinforced by the expiration of statutory deadlines under the previous law (taxing authority used to generate monies for the fund had expired on September 30, 1985). In a letter to congressional leaders, EPA Administrator Lee Thomas suggested that failure to reauthorize Superfund in a timely fashion would result in the cancellation of work contracts at cleanup sites and the furloughing of workers. According to Thomas, "While it will only take a few months to dismantle the program, it will take years and many millions of dollars to rebuild it."[69]

Progress on a compromise bill proceeded at a glacial pace because of sharp policy differences within the ranks of congressional Democrats as well as interchamber differences. One major obstacle was overcome in late July when conferees managed to reach an agreement covering all substantive policy issues except for the financial arrangements. Many of the changes were applauded by representatives of environmental organizations, who had worked hard to generate grass-roots pressure from communities throughout the United States on members of Congress to support the strengthening of Superfund.[70]

A consensus on funding was complicated by the opposition of public officials within the Reagan Administration, especially OMB administrators and Treasury Secretary James Baker. They were concerned about overall spending totals contained in the bill as well as the incorporation of new taxes on business to finance the program. However, conferees were ultimately persuaded that their own assessment of funding requirements revealed a need for sizable and stable revenue sources. The dispute over whether the financial burden should be borne by the private sector writ large or by petrochemical firms was settled by a decision to levy taxes on both. The conference bill passed both houses by an overwhelming margin and it was forwarded to President Reagan.

A question remained as to whether the president would choose to vent his dissatisfaction with the program funding arrangements by vetoing the bill. Superfund supporters sought to convey a pair of messages designed to highlight the political implications of a presidential signature or veto; these messages carried weight in part because of the overwhelming margins of victory for the bill in both houses of Congress, thus ensuring a veto override if necessary. First, congressional leaders—Republicans as well as Democrats—made it clear that they were willing to remain in session as long as it took to forestall the possibility of a pocket veto (a bill dies if the president does not sign it within ten days and Congress has since adjourned).[71]

Second, several Republican members of Congress, including Senator John Chafee (R–Rhode Island) and Senate Majority Leader Bob Dole (R–Kansas) sent a letter urging the president to sign the law to avoid saddling congressional Republicans with an unpopular campaign issue just prior to November elections.[72] In the end, President Reagan chose a pragmatic rather than an ideological course of action and signed the Superfund Amendments and Reauthorization Act (SARA) on October 20, 1986.

The reauthorization of Superfund was driven in part by statutory deadlines contained in the 1980 law and by the understandable desire to avoid the costs of stopping and restarting cleanup projects. In addition, a bipartisan coalition in Congress saw an opportunity for policy reform. This desire for change was prompted by dissatisfaction with the program management philosophy espoused by EPA administrators as well as new information indicating that the number of abandoned dump sites was larger than the legislators had realized in 1980. Political conflict, in this case, was less a function of partisan warfare than one of sharp policy disagreement between Congress and the president. Particularly disturbing to many members of Congress, as well as to the environmental policy community, was the inability or unwillingness of EPA officials to initiate permanent cleanup actions at NPL sites—as of 1985, only six sites had been cleaned up.

SARA, like HWSA, represented a fundamental policy shift from the original legislation in both scope and substance. Its $8.5-billion fund provided better than a fivefold increase in appropriations. Monies were derived from a tax on petroleum ($2.75 billion), a tax on chemical feedstocks ($1.4 billion), a broad-based business tax ($2.5 billion), general revenues ($1.25 billion), and interest plus cleanup expenses recovered from firms responsible for polluting an NPL site.[73] Industry officials had little to show for their efforts to shape the financial package except for the decision to delete the funding for the cleanup of oil spills. This suggests that Congress was becoming more sensitive to the claim that EPA was saddled with increasing policy responsibilities without a corresponding increase in staff or financial resources.[74]

The new law also set stricter standards and used goals and timetables as enforcement tools. EPA was instructed to begin work at 375 sites within five years. Appropriate criteria guiding site management decisions were also spelled out, including the use of permanent cleanup methods or detoxification in lieu of storing wastes within landfills. The intent of Congress to minimize or reduce industry dependence on land-based disposal was affirmed within SARA as well as HWSA.

Other notable features of SARA included a congressional effort to supplement EPA's regulatory responsibilities with an enforcement role for the public and private sectors. Citizens' lawsuits were authorized along with expanded procedural requirements for public comment and participation in proposed cleanup plans. Policymakers also made use of insurance and liability rules. Owners of underground tanks storing petroleum or other chemical pollutants were required to maintain sufficient financial reserves or insurance to cover damages stemming from accidental leaks.

Finally, a section of the law gave the federal government the authority to override state statutes of limitations in permitting citizens to sue for injury caused by exposure to hazardous wastes. The chief argument for this provision was based on the belief that citizens needed to preserve their right to sue polluting firms since a number of environmentally induced illnesses (such as some forms of cancer and several respiratory ailments) might not emerge for an extended period of time.

CONCLUSIONS

An examination of congressional response to RCRA and Superfund policy proposals illustrates the use of different approaches to agenda setting and adoption. The decisions to consider and authorize RCRA are more representative of Kingdon's characterization of "hidden specialists" within government. The congressional agenda was not reached through interest-group channels or presidential messages. EPA officials and staffers within the environmental policy committees folded the hazardous waste regulatory program into the larger solid-waste management policy.

In addition, the research and development phase preceded the call for regulatory action in much the same way that air- and water-quality policy issues had evolved. However, a low-key approach was necessitated by the realization that federal environmental policies in 1975 could not command the same type of unswerving support as pollution-control policies of the early 1970s. RCRA architects were aware of competing policy concerns such as energy development and regulatory compliance costs. Their task was also facilitated by the disproportionate attention given by political activists to congressional deliberations over TSCA, a policy proposal that produced considerable anxiety within the boardrooms of chemical firms.

On the other hand, the need for a national hazardous waste policy to deal with the cleanup of abandoned dump sites was recognized by members of Congress because of media coverage of Love Canal and related incidents. This led to a rise in public concern, increased activity by environmental organizations, and a decision by EPA officials to recommend a policy solution based on public health rather than environmental quality goals. Superfund proposals were thus propelled onto the institutional agenda by insiders within EPA and key members of Congress, with a boost from the press and environmental organizations. Superfund's subsequent adoption was complicated by industry opposition and inaction within the Senate Finance Committee but was ultimately approved because of the efforts of key legislators such as Representative Florio and Senator Stafford, the interdependent effects of public opinion and media attention, and, to a degree, the widely shared belief that a similar proposal would have little or no chance of passage after Ronald Reagan became president.

The reauthorization of RCRA and Superfund was accompanied by a significant increase in resources and programmatic responsibilities. In each case, congressional action was fueled by media attention to political turmoil within EPA or to environmental catastrophes. Policy change reflected not only a willingness in Congress to incorporate new information about hazardous waste but a lack of trust in the Reagan Administration's commitment to program management.

Congress as an institution managed to fill a vacuum within the hazardous waste policy realm in the early 1980s. Because the Reagan Administration was committed to the decentralization and deregulation of domestic federal programs, its participation in the extension of existing regulatory policies was limited. Thus, Congress rather than the presidency took the lead in reauthorizing toxic waste legislation. Moreover, this shift did not represent a significant alteration in the strength of competing interest-group coalitions or partisan advantage. We suggest that the critical factors influencing Congress included an inclination to accept technical information on hazardous waste from the Office of Technology Assessment and the National Academy of Sciences rather than EPA[75] and its sensitivity to the strength and consistency of public support for hazardous waste policies.

NOTES

1. Robert Eyestone, *From Social Issues to Public Policy* (New York: John Wiley & Sons, 1978), p. 3.

2. Roger W. Cobb and Charles D. Elder, *Participation in American Politics: The Dynamics of Agenda Setting* (Boston: Allyn & Bacon, 1972), p. 85.

3. Cobb and Elder, p. 96.

4. An extended discussion of the grass-roots model can be found in Roger Cobb, Jennie Keith-Ross, and Marc Howard Ross, "Agenda Building as a Comparative Political Process," *American Political Science Review, 70* (March 1976).

5. John W. Kingdon, *Agendas, Alternatives, and Public Policies* (Boston: Little, Brown, 1984), pp. 72–74.

6. Kingdon, p. 174.

7. Cited in Mary Etta Cook and Roger H. Davidson, "Deferral Politics: Congressional Decision Making on Environmental Issues in the 1980s." In Helen Ingram and Kenneth Godwin, eds., *Public Policy and the Natural Environment* (New York: JAI Press, 1985).

8. Charles O. Jones, *Clean Air: The Policies and Politics of Pollution Control* (Pittsburgh: University of Pittsburgh Press, 1975).

9. Alfred A. Marcus, *Promise and Performance: Choosing and Implementing an Environmental Policy* (Westport, Conn.: Greenwood Press, 1980), especially pp. 63–64.

10. David Schnapf, "State Hazardous Waste Programs under the Federal Resource Conservation and Recovery Act," *Environmental Law, 12* (Spring 1982), p. 684.

11. U.S. Environmental Protection Agency, *Disposal of Hazardous Wastes*. Report to Congress submitted by the EPA Office of Solid Waste Management Programs, 1974 (hereafter referred to as the EPA Report).

12. EPA Report, p. 13.

13. Schapf, pp. 684–85.

14. Samuel Epstein, Lester Brown, and Carl Pope, *Hazardous Waste in America* (San Francisco: Sierra Club Books, 1982), p. 190.

15. Kingdon, pp. 72–74.

16. Kingdon, pp. 130–31.

17. Gary Dietrich, "Ultimate Disposal of Hazardous Wastes." In Robert J. Pojasek, ed., *Toxic and Hazardous Waste Disposal, VIII* (Ann Arbor, Mich.: Ann Arbor Science, 1980), pp. 1–11.

18. U.S. Office of Technology Assessment, *Technologies and Management Strategies for Hazardous Waste Control* (Washington, D.C.: Government Printing Office, 1983), p. 8 (hereafter referred to as the OTA Report).

19. OTA Report, p. 8.

20. OTA Report, pp. 3–40.

21. Riley E. Dunlap, "Public Opinion and Environmental Policy." In James P. Lester, ed., *Environmental Politics and Policy: Theories and Evidence* (Durham, N.C.: Duke University Press, 1989), p. 161.

22. Cobb and Elder, p. 83.

23. Adeline Levine, *Love Canal: Science, Politics and People* (Lexington, Mass.: Lexington Books, 1982).

24. Levine, Chapter 7.

25. Epstein, Brown, and Pope, Chapter 1.

26. Marc Landy, "Ticking Time Bombs!!! EPA and the Formulation of Superfund." In Helen Ingram and Kenneth Godwin, eds., *Public Policy and the Natural Environment* (New York: JAI Press, 1985), p. 246.

27. Much of their influence can be attributed to a combination of personal interest and institutional position. Muskie headed an environmental pollution subcommittee within the Senate Public Works Committee for several years, while Waxman continues to chair a subcommittee dealing with health and the environment within the House Energy and Commerce Committee (as of 1992).

28. Cobb and Elder, Chapter 8.

29. James E. Anderson, *Public Policy-Making* (New York: Praeger, 1975), p. 66.

30. John W. Kingdon, *Congressmen's Voting Decisions*, 2nd ed. (New York: Harper & Row, 1981).

31. Representative Paul Rogers, "Solid Waste Disposal Act Extension—1974." Opening remarks before the Subcommittee on Public Health and the Environment,

Committee on Interstate and Foreign Commerce, U.S. House of Representatives, March 27, 1974, p. 1.

32. Rogers, opening remarks, p. 2.

33. Rogers, opening remarks, p. 2.

34. Russell Train, Administrator, Environmental Protection Agency, "Solid Waste Disposal Act Extention—1974." Testimony before the Subcommittee on Public Health and the Environment, Committee on Interstate and Foreign Commerce, U.S. House of Representatives, March 27, 1974, p. 143.

35. CQ Almanac, "Solid Waste Programs" (Washington, D.C.: Congressional Quarterly, 1974), p. 835.

36. Epstein, Brown, and Pope, p. 190.

37. Cook and Davidson, pp. 56–57.

38. Public Law No. 94-580, Section 3002.

39. Cited in Epstein, Brown, and Pope, p. 206.

40. John F. Mahon, "Corporate Political Strategies: An Empirical Study of Chemical Firm Responses to Superfund Legislation." In *Research in Corporate Social Performance and Policy, 5* (New York: JAI Press, 1983).

41. Mahon, p. 155.

42. Landy, p. 250.

43. Landy, p. 251.

44. Epstein, Brown, and Pope, pp. 208–10.

45. George Clemon Freeman, Jr., *The Superfund Section 301(e) Study Group Recommendations and Related Funding Proposals*. Report prepared for the National Conference on Environmental Injury Compensation, American Insurance Association, Washington, D.C.: March 1984.

46. Mahon, pp. 160–61.

47. Epstein, Brown, and Pope, p. 212.

48. Mahon, p. 167.

49. Kathy Koch, "Compromise Reaching on 'Superfund' Bill," *Congressional Quarterly*, November 29, 1980, pp. 3435–37.

50. Mahon, pp. 172–73.

51. Anderson, pp. 93–95.

52. Paul A. Sabatier, "Knowledge, Policy-Oriented Learning, and Policy Change," *Knowledge: Creation, Diffusion, Utilization, 8* (June 1987), pp. 649–92.

53. Anthony Downs, "Up and Down with Ecology—The 'Issue–Attention' Cycle," *The Public Interest, 28* (Summer 1972), pp. 38–50.

54. Michael E. Kraft, "The Political and Institutional Setting for Risk Analysis." In Vincent Covello, Joshua Menkes, and Jeryl Mumpower, eds., *Evaluation and Management* (New York: Plenium Press, 1986).

55. These concerns are summarized in a pair of studies by the General Accounting Office. See *Waste Disposal Practices: A Threat to Health and the Nation's Water Supply*, June 16, 1978, and *How to Dispose of Hazardous Waste: A Serious Question That Needs to Be Resolved*, December 19, 1978.

56. Richard Riley, "Toxic Substances, Hazardous Wastes and Public Policy: Problems in Implementation." In James P. Lester and Ann O'M. Bowman, eds., *The Politics of Hazardous Waste Management* (Durham, N.C.: Duke University Press, 1983), p. 36.

57. OTA Report, op. cit.

58. National Research Council, *Management of Hazardous Industrial Waste: Research and Development Needs* (Washington, D.C.: National Academy Press, 1983).

59. Richard Barke, "Policy Learning and the Evolution of Federal Hazardous Waste Policy, " *Policy Studies Journal, 14* (September 1985), p. 127.

60. The Times Beach controversy is discussed within the context of risk perception and the responsibilities of public officials by Walter Rosenbaum. See *Environmental Politics and Policy* (Washington, D.C.: CQ Press, 1985), pp. 79–81.

61. Michael Reese, "Storm Over the Environment," *Newsweek*, March 7, 1983, pp. 16–19.

62. CQ Almanac, "Congress Tightens Hazardous Waste Controls" (Washington, D.C.: Congressional Quarterly, 1984), p. 305.

63. 1984 CQ Almanac, p. 307.

64. 1984 CQ Almanac, p. 308.

65. Robert Cameron Mitchell, "Public Opinion and Environmental Politics in the 1970s and 1980s." In Norman Vig and Michael Kraft, eds., *Environmental Policy in the 1980s: Reagan's New Agenda* (Washington, D.C.: CQ Press, 1984), pp. 55, 57.

66. General Account Office, *Inspection, Enforcement and Permitting Activities at New Jersey and Tennessee Hazardous Waste Facilities*. See also Charles Davis, "Implementing the Resource Conservation and Recovery Act of 1976: Problems and Prospects, " *Public Administration Quarterly* (Summer 1985), pp. 218–36.

67. CQ Almanac, "House, Senate Pass Superfund Authorization" (Washington, D.C.: Congressional Quarterly, 1985), p. 191.

68. 1985 CQ Almanac, p. 192.

69. CQ Almanac, "Reagan Signs 'Superfund' Waste Cleanup Bill" (Washington, D.C.: Congressional Quarterly), 1986.

70. Phillip Shabecoff, "Environmentalists Turn Over a New Leaf, Sort Of," *The New York Times*, October 26, 1986.

71. 1986 CQ Almanac, p. 111.

72. 1986 CQ Almanac, p. 111.

73. 1986 CQ Almanac, p. 112.

74. Rosenbaum argues that the imbalance between program responsibilities and resources is troublesome for federal pollution-control programs in general. See Walter Rosenbaum, *Environmental Politics and Policy* (Washington, D.C.: CQ Press, 1985).

75. Barke, p. 130.

3

SHAPING HAZARDOUS WASTE POLICY DECISIONS AT EPA

The adoption of federal hazardous waste laws by Congress represents an important achievement within the realm of environmental protection policymaking. It signifies that lawmakers have added statutory protection against land-based pollutants to the existing arsenal of air and water pollution-control policies. However, statutory protection taken alone is meaningless unless steps are taken to develop and implement program guidelines by public agencies with requisite amounts of financial and staff resources, political support, and enforcement authority. EPA clearly plays a pivotal role in this process as the agency designated by Congress to carry out hazardous waste policies.

This chapter focuses on the decision-making environment of EPA as an implementing agency. Of particular concern are the statutory, political, organizational, and fiscal factors that shape programmatic choices. These factors occasionally reflect tension between agency goals and those of other governmental institutions inside and outside the executive branch whose operations are affected by RCRA and Superfund requirements. The resulting political climate can either simplify or complicate enforcement efforts.

DECISION-MAKING ENVIRONMENT

EPA officials work in a political environment that includes a number of policy actors with the ability to influence or resist agency decisions. Statutory language provides the most visible example of institutional impact. Laws vary along such criteria as the clarity of policy goals, the degree of hierarchical integration within and among implementing institutions, and the presence or absence of opportunities for participation in the decision-making process by citizens and interest groups. Second, the administrative setting reflects the importance of organizational characteristics—both formal and informal. This includes not only a discussion of who is responsible for what in a legalistic sense but an appreciation for the diversity of policy preferences held by individuals within the agency as well.

Third, the political context directs attention to efforts made by other agencies, institutions, and interest groups to affect hazardous waste policy decisions. A fourth and particularly critical factor is the resource base. Any regulatory agency requires money and staff to administer programmatic activities such as research and development, rulemaking, the issuance of permits, and monitoring and enforcement. In a collective sense, these factors narrow the range of decision-making options available to EPA officials.

POLICY CONTEXT

Members of Congress are concerned about the amount of discretion wielded by administrators, particularly those responsible for the management of regulatory programs. Environmental protection policies often include areas of decision making such as standard setting or approaches to meeting these standards that are typically reserved for experienced administrators in other areas of policy. For example, the language of the Clean Air Act reflects a conscious effort to incorporate precise goals for reducing the volume of airborne pollutants and timetables for their attainment in order to reduce the probability of agency capture by regulated industries.[1]

In effect, the architects of this policy were expressing their skepticism about the ability or willingness of agency officials to resist industry pressures for the delay and/or dilution of environmental quality standards and their enforcement. As a result, EPA officials are occasionally unable to recommend a course of action based on professional judgment because of legislative prescription of actions to be taken in response to a given problem or situation.

Like other environmental protection policies, RCRA and Superfund place limits on discretionary authority to be exercised by EPA officials. However, the adoption of these laws took place in the latter part of the 1970s; hence, policymakers were operating in a political atmosphere that was sensitive to economic as well as environmental values. In addition, hazardous waste as a policy problem was understood less well (in the sense of knowing the relative effectiveness and cost of policy options) than air and water pollution problems. Thus, policy goals expressed within RCRA and Superfund were less ambitious than either the Clean Air Act or the Clean Water Act in identifying deadlines and performance targets for waste reduction and the number of dump sites to be cleaned up.

Congressional dissatisfaction with the pace of hazardous waste program implementation in the early 1980s led policymakers to consider a more directive approach. A goals and timetables approach was viewed as one way of accelerating EPA efforts to achieve policy objectives.[2] The Hazardous and Solid Waste Amendments of 1984 incorporated the so-called "hammer clause," which posed the threat of a ban on the disposal of selected wastes or the automatic adoption of the California list of hazardous wastes if EPA failed to take action on these matters by a given date.

In like fashion, the Superfund Amendments and Reauthorization Act of 1986 (SARA) called for EPA to inspect and evaluate each facility on its list of the most hazardous dump sites within four years or to provide a reasonable explanation of why it was unable to do so. SARA also required schedules for the completion of feasibility studies for sites on the National Priorities List (NPL) and for the beginning of remedial cleanup actions.

Hazardous waste statutes also identify the form and direction of intergovernmental relationships affecting program management decisions. RCRA is one of several pollution-control policies adopted in the 1970s that is based on the "partial preemption" approach.[3] A key feature of this approach is the involvement of both federal and state environmental agencies in a collaborative effort to reduce, recycle, or eliminate chemical pollutants.

The first step is the establishment of minimum environmental quality standards for the treatment, storage, or disposal of hazardous wastes by EPA, which are uniformly applicable to regulated firms and governments within the United States. State officials are then encouraged to submit plans demonstrating their willingness and ability to take over program management responsibilities. Authority is delegated to state policymakers if the policies or regulations submitted for review are at least as stringent as those of the federal government. In addition to their acquisition of greater decision-making autonomy, states become eligible for the receipt of federal funds providing a portion of RCRA operating costs.

Unlike other preemptive policies such as the Clean Air Act, RCRA allows state involvement to be a matter of choice. If state officials choose not to participate, the management of RCRA is undertaken by EPA administrators operating out of the regional office. Federal preemption can occur if state policymakers violate the terms of the agreement under which program authority is assigned. For example, EPA might choose to assume program management responsibilities if the state agency failed to hire a sufficient number of inspectors or to carry out enforcement actions against polluting firms. In short, the partial preemption policy design allows a balance of sorts between federal and state control over policy decisions.

The Superfund program does not incorporate a preemptive policy design but it does offer significant decision-making responsibilities to multiple political actors. A formal agreement must be developed and approved by EPA and the affected state government for each NPL site. This applies to both forms of enforcement—emergency actions aimed at ameliorating immediate threats to public health or environmental quality and permanent cleanup of abandoned dump sites. Lengthy intergovernmental negotiations may follow since these agreements spell out the proportion of cleanup costs to be borne by the state. Another outcome is the designation of a lead agency for site operations. If EPA is in charge, the work is contracted out to private-sector firms and state officials serve as project advisers. If the state assumes responsibility, administrators are asked to develop a work plan and to hire cleanup personnel, while EPA's role shifts to one of oversight.

Yet another way in which policy implementation is affected by statutory construction is the insertion of citizen participation opportunities within the decision-making process. In some cases, legislators will yield to the objections of regulated industry representatives and omit or resist public input. Both RCRA and Superfund initially contained provisions for public comment but provided few statutory toeholds for citizens to obtain legal redress of grievances in court.[4]

This rather limited approach provided a stark contrast to the opportunities for public involvement found in other statutes. Several environmental policies allow public input to be considered in several phases of decision making, thereby maximizing one's chances of exercising influence. For example, the federal surface mining program calls for public comment to take place before a permit is issued to a coal company as well as during the review process for proposed regulations. Types of participation (testimony at an administrative hearing, lawsuits), the extent of agency outreach to individuals or organizations with a stake in the decision, and the willingness of administrators to incorporate citizen input into the final decision vary as well.[5]

Congress acted to strengthen public participation opportunities in the reauthorized versions of RCRA and Superfund during the 1980s. Under RCRA, an ombudsman's office has been created within EPA to handle citizen complaints and questions about program operations. In addition, EPA is now required to provide ample opportunity for public notice and comment before entering into any settlement with one or more firms over actions to be taken in dealing with an imminent hazard. Citizen lawsuits are also authorized as a means of enforcing the ban on open dumping of hazardous waste.[6]

Similar changes have been incorporated within the Superfund program. Litigation as a means of combating EPA nonenforcement is an option now available to individuals or organizations. However, in one important respect, the law extends public participation beyond prior statutory efforts. Congress added an expanded interpretation of liability allowing the federal government to override state time limits that would otherwise restrict the ability of citizens to sue for injury allegedly caused by exposure to hazardous waste.

EPA's role is both enhanced and limited by these participatory developments. Decisions made by EPA officials can reflect information based on a greater diversity of viewpoints. This, in turn, adds an extra dose of legitimacy to agency actions. On the other hand, public involvement decreases the likelihood that EPA administrators can operate quietly behind the scenes with minimum scrutiny from the public, media sources, or Congress.

ADMINISTRATIVE CONTEXT

It is impossible to analyze the role of EPA in shaping the implementation of hazardous waste programs without an understanding of how it emerged, organizational attributes, and the degree of commitment to agency objectives shown by agency administrators. President Richard Nixon issued an executive order creating EPA in 1970. Nearly 6,000 employees from 15 pollution-control programs scattered throughout the federal bureaucracy were brought under the jurisdiction of a new, independent agency. Included were a number of rather sizable agencies dealing with air pollution (from the Health, Education, and Welfare Department) and water pollution (from the Interior Department).[7]

At the top of the organizational hierarchy (see Figure 3–1) is an Administrator, and deputy administrator who serve at the pleasure of the president. The decision to employ a single executive model rather than the more commonly used commission model makes EPA unique among federal regulatory agencies in that responsiveness and perfor-

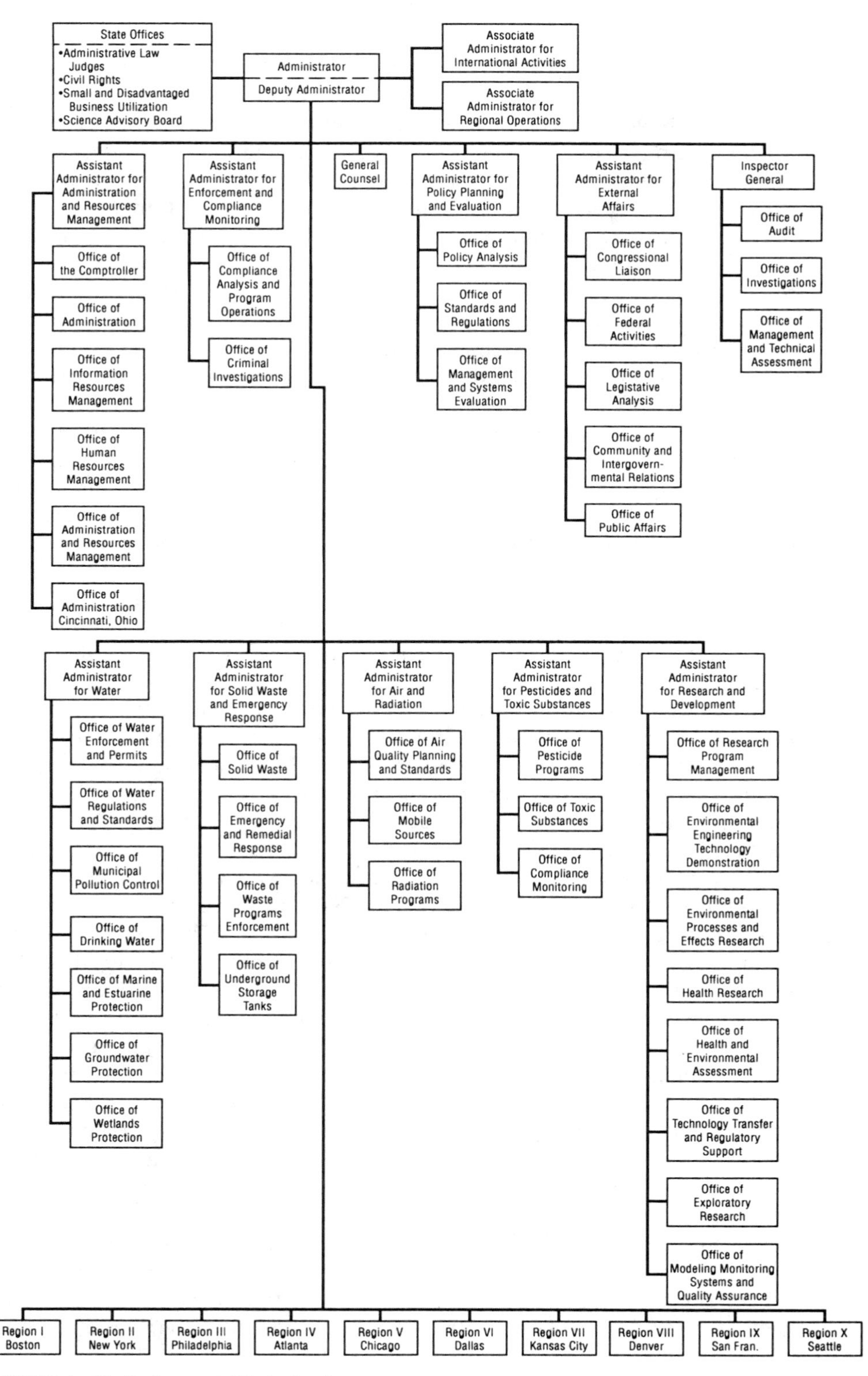

FIGURE 3–1 The Environmental Protection Agency

Source: U.S. Government Manual (Washington, D.C.: Government Printing Office, 1990), p. 554.

mance are given higher priority than political independence from the president and Congress.[8] This is reflected in the degree of formal and informal authority placed in the administrator's hands. He or she has final say over rulemaking initiatives, is responsible for budget preparation, and has considerable latitude to reorganize agency operations. The administrator's authority also is enhanced by the technical uncertainty associated with the resolution of pollution problems. Risk-based decisions are justified more easily since a wide range of consequences will inevitably accompany any course of action under consideration.[9]

This is not to say that EPA leaders can direct and control the behavior of subordinates without considerable effort. The administrator's ability to resolve within-agency disputes often depends less on the exercise of formal authority than on negotiating skills. Conflict management is complicated by a rather cumbersome set of organizational relationships within EPA that simultaneously embrace an integrative approach to environmental management and a programmatic orientation. On the one hand, functional management categories have been adopted in agency offices representing planning and management, standards and compliance, and research.[10] This approach is presumably useful as a means of examining environmental problems in an interrelated fashion.

On the other hand, program categories are also maintained. This is a legacy of the original reorganization plan, which shifted both programs and personnel to the newly created EPA from other departments. Administrators heading these programs were not prepared to make the transition from a single-issue focus to a more holistic view of environmental management. To undertake such an effort would have been an invitation to chaos—at least in the short run. William Ruckelshaus and his advisers chose to leave the programmatic units intact to ensure that the fledgling agency could "hit the ground running."[11] However, subsequent analyses of EPA behavior reveal the problem of interoffice disputes between assistant administrators, which are attributed to the uneasy coexistence of various administrative units dominated by a policy perspective, a research perspective, or a program perspective.[12]

Hazardous waste programs have been assigned to the Office of Solid Waste and Emergency Response (OSWER) within EPA. Like other environmental protection programs, OSWER is headed by an assistant administrator whose responsibilities include the development of policies, standards, and regulations for RCRA and Superfund; the enforcement of these laws; and the provision of technical assistance on waste management operations to other federal, state, and local governments.[13] Program staff are also found in each of EPA's ten regional offices throughout the United States. This is the principal point of contact with the federal government for state and local environmental officials as well as industry representatives. One concern to be discussed more fully in a later section is the inconsistency found in the enforcement of hazardous waste programs across federal regions.[14]

Important administrative units contained within OSWER include the Office of Emergency and Remedial Response, the Office of Waste Programs Enforcement, and—since 1984—the Office of Underground Storage Tanks. Administrators within OSWER maintain a programmatic perspective; that is, emphasis is placed on implementing statutory objectives. However, decision-making recommendations must be cleared with other EPA actors inside the Office of Standards and Regulations or the Office of Policy Analysis who will raise questions about the technical adequacy of a prospective course of action or its cost.

While organizational arrangements can occasionally impede or delay program management decisions, a particularly crucial ingredient for a successful administrative enterprise is the commitment of key administrators to hazardous waste policy objectives. The easiest way for a newly elected president to demonstrate support for environmental protection policy goals (at least in the short run) has been for him to select an EPA administrator with a proven track record in the management of pollution-control programs and the belief that government can make a positive contribution toward the resolution of policy problems.

Ronald Reagan's predecessors in the Oval Office were supportive of EPA and policy initiatives dealing with environmental protection. President's Nixon and Ford were somewhat more sensitive to the economic ramifications of environmental regulation than was Jimmy Carter; however, no one questioned the basic rationale for federal pollution-control programs. Each recommended increases in EPA's budget and staff to reflect its expanding programmatic responsibilities and appointed an administrator with previous experience in environmental management.[15]

The election of President Reagan in 1980 resulted in the subordination of environmental policy objectives to economic growth and deregulation. These priorities were faithfully reflected by EPA Administrator Anne Gorsuch (later Burford) and Rita Lavelle, the Assistant Administrator for Hazardous Waste Programs. Neither appointee was particularly knowledgeable about pollution control problems; neither had environmental management experience. However, each demonstrated a willingness to make program decisions within the context of Reagan's larger policy agenda.[16] An early indicator of their loyalty was the submission of a budget request calling for a sizable decrease in funding and staff.[17]

A more tangible example of the relationship between organizational commitment and program operations can be shown by comparing Superfund decision making under the latter stages of the Carter

Administration with the first two years of the Reagan Administration. According to Cohen, Carter's view of Superfund was captured by the expression, "shovels first, lawyers later."[18] EPA administrators were actively working on a management plan for the cleanup of abandoned dump sites prior to the formal adoption of Superfund. This early action was achieved by using existing authority from Section 311 of the federal Water Pollution Control Act to create a small emergency response program. A task force consisting of personnel drawn from RCRA and Clean Water programs also was established to gear up for the eventual prosecution of companies, individuals, or governmental jurisdictions for the illegal dumping of hazardous waste. In short, administrators were prepared to take action, given the availability of legal authority and increased funding.

Under Reagan, EPA's approach to the implementation of Superfund changed markedly. Inertia replaced action because of a philosophical shift in the pace and direction of enforcement. Both Burford and Lavelle favored a nonconfrontational approach over litigation as a means of dealing with regulated groups. They believed that industry would cooperate with government and that reliance on lawsuits would only serve to alienate corporate officials and further delay the cleanup of contaminated sites. Accordingly, results would be maximized by EPA-industry negotiations over the extent of corporate liability on a given site with minimum or no opportunity for comment from public or environmental organizations.[19] Emphasis was placed on a different set of decisional criteria—namely, the maintenance of cordial relationships with industry officials and the minimization of governmental expenditures for cleanup actions.

Yet another manifestation of organizational commitment is the promulgation of decision rules. Agency administrators can influence the probability of industry or state governmental compliance with statutory objectives by developing rules that specify the boundaries of permissible action. Shortly after the adoption of Superfund, EPA put forward a regulation requiring states to contribute 10 percent of the initial planning and design costs at a site in addition to the statutorily mandated cleanup costs. According to Bowman, the practical effect of this rule was to constrain state governments at the front end, since many did not budget their own resources for planning work.[20] Moreover, cleanup activities by law could not begin without a feasibility study. After public officials became aware of this predicament, state financial participation in the design and planning phase was amended to require full federal funding.

The foregoing example illustrates a commonly voiced concern in the evaluation of prospective rules by EPA officials. Administrators occasionally fail to consider the possibility that regulatory actions may have unintended consequences for the attainment of environmental quality objectives or the distribution of market impacts. In addition, the cumulative impact of these regulations clearly affects some industries more than others. Smaller firms operating on the margins of profitability find it difficult to comply with newly imposed insurance and/or reporting requirements mandated under HSWA.[21] For many, the options are both stark and unpleasant—cut costs by refusing to comply with regulatory decisions or go out of business while trying to obey the law.

POLITICAL CONTEXT

Administering public programs through an intergovernmental maze of elected officials and administrators is a complex undertaking that involves numerous checkpoints or decision clearances along the way. EPA must deal with numerous policy actors within the executive branch, Congress, the courts, and state and local government as well as organized interests and the media. Efforts to influence EPA's management of hazardous waste programs are directed toward specific administrative decisions (such as rulemaking or enforcement) or the allocation of resources to RCRA and Superfund (discussed in the following section).

Key EPA contacts within the executive branch include the president, staff agencies such as the OMB and the OTA, and other departments, notably the Department of Defense (DOD), the Department of Energy (DOE), and the Department of the Interior (DOI).

Presidency

Direct presidential involvement rarely occurs within the EPA. Institutional impact is more likely to be manifested in the appointment of the EPA administrator and deputy administrator and the willingness of the president to accept or tolerate their decisions in the face of pressure from organized interests and staff agencies. One such example is President Reagan's acquiescence to an EPA buyout plan for Times Beach, Missouri, as the preferred means of dealing with dioxin contamination.

Office of Management and Budget

Perhaps the major institutional obstacle for EPA within the executive branch is the OMB. Historically, departments have found it necessary to deal with OMB as the key staff agency with responsibilities for legislative clearance of policy proposals plus review of budget requests and personnel ceilings.[22] Control is vested within the realm of budgetary execution as well. OMB staffers monitor the pace of agency expenditures throughout the fiscal year. On occasion, this monitoring adversely affects agency flexibility to carry out program activities. For example, EPA's request for supplemental funds in fiscal year 1985 to implement the 1984 RCRA amendments was turned down by OMB, which, in turn, directed agency administrators to find the money from existing program resources.

A more critical source of tension stems from OMB's growing policy role in the review of agency regulations.[23] This began in the early 1970s with President Nixon's Quality of Life review process, which was placed under the direction of OMB Director George Schultz. Other agencies with an interest in regulations proposed by EPA were invited to review preliminary drafts with an eye toward industry compliance costs and to offer comments. Key concerns included two rather basic questions: Is this rule necessary? If so, can it be fine-tuned in a way that provides a balance between economic health and environmental quality?

President Ford added a requirement that an "inflation impact statement" be appended to proposed rules so that OMB and other agencies could better assess their economic consequences. Yet another variation on the regulatory review theme was incorporated by President Carter, who created an interagency council referred to as the Regulatory Analysis Review Group (RARG). This group also employed the services of economists from the Council of Economic Advisors, whose primary responsibility was to prepare cost-benefit analyses of major regulations (i.e., rules estimated to equal or exceed $100 million in compliance costs). These analyses became part of the rulemaking record but were rarely (if ever) used to unilaterally kill a proposed rule.

These precedents were expanded further by President Reagan with the aid of OMB Directors David Stockman and James Miller. Oversight of agency rulemaking was significantly strengthened by provisions of the Paperwork Reduction Act of 1980, which created the Office of Information and Regulatory Affairs (OIRA) within OMB, and Reagan's issuance of Executive Order 12291, which required federal agencies to submit all proposed regulations to OMB for review. Each rule packet had to include a Regulatory Impact Analysis (RIA), which provided documentation of alternatives and costs plus explanations for the exclusion of alternatives not chosen.[24] The effect of these changes on OMB operations was aptly summarized by former Deputy Director Jim Tozzi, who declared, "The Government works using three things: money, people and regulations; the agency must get all three through OMB."[25]

In 1985, President Reagan issued Executive Order 12498, which extended OMB review processes to include front-end decision making as well. Materials to be submitted to OIRA now included agency regulatory policies, objectives, and rulemaking plans for the forthcoming year, which could be evaluated in view of the president's legislative priorities. This directive in combination with the earlier order immediately drew fire from critics on Capitol Hill, who charged that OMB was usurping Congress's policymaking prerogatives. On the other hand, defenders of centralized regulatory authority argued that these actions enabled OMB as the president's agent to determine whether a proposed rule contained sufficient analytical support to justify the conclusion that the public was receiving benefits from the selection of a "least cost to society" regulatory option.[26]

Has the increase in OMB's oversight responsibility produced a chilling effect on EPA's rulemaking activities? Data presented within OMB's annual regulatory report indicate that EPA is less likely than other agencies to submit regulatory proposals that pass OMB muster without changes. Approximately half are held up for detailed scrutiny, compared with an average of 15 percent for other departments and independent agencies. In addition, the amount of delay associated with OMB review of major regulatory proposals forwarded by EPA has steadily increased between 1986 and 1989. One consequence has been a decline in the absolute number of rules produced.

On the other hand, the ability of OMB to veto major regulatory proposals or to require major changes has been limited by political maneuvering undertaken by EPA and its allies in Congress. An end-run approach is occasionally used at the information-gathering stage of regulatory development by acquiring data from contractors or consultants rather than from surveys (which must be approved by OMB under the Paperwork Reduction Act).[27]

Another tactic used by EPA insiders to deal with OIRA analysts is to elevate the visibility of regulatory concerns by discreetly contacting members of Congress or the media. In 1985, EPA proposed a rule calling for an eventual ban on all uses of asbestos, a known carcinogen. OMB officials responded by recommending the transfer of regulatory authority from EPA to the Labor Department's Occupational Safety and Health Administration (OSHA), which had more limited authority over the production of the substance. EPA staffers leaked details of the proposed transfer to Representative John Dingell (D–Michigan), who subsequently called a committee meeting to investigate the matter. OMB officials reconsidered, and by 1986 EPA had reasserted its authority over asbestos.[28]

Finally, the inclination of OMB administrators to delay certain types of proposed regulations was dealt a blow in 1986 by a decision rendered in the Washington, D.C., District Court.[29] The case revolved around a lawsuit filed by the Environmental Defense Fund (EDF) in

1985 against EPA Administrator Lee Thomas for missing the statutory deadline of March 1 for the issuance of rules pertaining to the storage of hazardous wastes in tanks. EDF also sued OMB, claiming that its regulatory review process could not be used in a way that resulted in an agency's inability to meet these deadlines. The court decision concluded that "OMB has no authority to use its regulatory review under Executive Order 12291 to delay promulgation of EPA regulations...beyond the date of a statutory deadline."[30]

Other Executive Agencies

Another problem originating within the executive branch involves getting federal departments to comply with RCRA and Superfund. The magnitude of land or water contamination from agency sources is serious, as our discussion in Chapter 1 indicates, and policymakers have taken a number of tentative steps to address these concerns. In 1978, President Carter issued Executive Order 12088, requiring all department heads to take whatever actions were necessary to prevent, control and abate pollution occurring at federal facilities.[31] Since then, more specific guidelines have been incorporated within the body of hazardous waste statutes. In addition, various administrators have initiated departmentwide actions to alleviate pollution problems and have, in some cases, signed a memorandum of agreement with EPA to undertake corrective action.

Unfortunately, it is often difficult to reconcile environmental protection goals with other substantive policy goals central to the main mission of a department. While examples of lax management practices can be found in many agencies, the lackluster performance of the Departments of Defense (DOD) and Energy (DOE) provides compelling evidence that enforcement efforts should not be confined to regulated industries. Until recently, DOE strongly resisted efforts by EPA to obtain data. Hazardous wastes generated as a byproduct of nuclear weapons production were often commingled with radioactive wastes. EPA's objection resulted in DOE's argument that a secrecy exemption contained in the Atomic Energy Act was still valid and that defense-related goals ought to take precedence over site-specific environmental problems.[32] EPA and state regulatory agencies have had little success in using negotiation as a means of producing voluntary compliance and have raised the stakes by issuing fines and penalties. However, DOE officials have resisted these pressures as well, claiming that sovereign immunity protects them from enforcement actions of this sort.

The General Accounting Office (GAO) has been quite critical of the military for its failure to provide adequate levels of environmental protection for its base employees and residents of adjacent communities.[33]

Efforts by EPA and state officials alike to coordinate data gathering, inspection, or cleanup activities have met with considerable resistance from base commanders.[34] On occasion, these decisions have produced adverse ecological impacts that could have been prevented had prior consultation with individuals more familiar with local conditions taken place. A case in point is a long-standing controversy involving the Rocky Mountain Arsenal (RMA), a weapons plant operated by the U.S. Army near Denver, Colorado.

Problems of off-base contamination have plagued RMA operations since the early 1950s, when nearby farmers reported evidence of crop damage. Studies subsequently conducted by university and government scientists confirmed that damages could be attributed to groundwater pollution migrating from the facility. Three other incidents have been documented, including the contamination of well water in Adams County in 1957, and the discovery of toxic chemicals in underground aquifers and surface water outside RMA boundaries in 1974 and again in 1981. The Colorado Department of Health (CDH) and EPA both attempted to force the Army to change its ways—to halt current releases of hazardous wastes, clean up contaminated areas, and establish a monitoring program to detect and mitigate future spills.[35]

Base officials eventually acknowledged that the Army bore some responsibility for mitigating the effects of their managerial mishaps. Following the 1974 incident, they decided to solve the problem by containing existing wastes. This course of action was clearly unacceptable to the residents of nearby communities and led to increased pressure from state and federal environmental officials as well as area members of Congress to adopt a more permanent solution. In December 1983 CDH filed a lawsuit against Shell Oil Company and the U.S. Army (under a provision of Superfund allowing third-party lawsuits for damage to natural resources) and requested that a cleanup remedy be pursued in lieu of containment. EPA followed up with a decision in October 1984 to place contaminated lands on the NPL.[36]

These actions in concert appear to have produced the desired effect. Later that month, a plan was released by Army authorities calling for the selection of the total cleanup option. While the operational phase is scheduled to continue well into the future, this particular case illustrates both the predictable resistance to environmental initiatives from agencies unaccustomed to acting in an ecologically responsible manner and the potential usefulness of combining the efforts of regulatory bureaus at differing levels of government and elected officials in pushing for compliance with pollution control laws.

A more complex example of interdepartmental politics at the federal level revolves around the issue of hazardous waste exports to developing countries. Important organizational values are at stake here, and they have produced bureaucratic warfare in the past between EPA and the Departments of State, Commerce, and Agriculture, among others.[37] Key policy questions are twofold: Should pollution-generating firms be allowed to circumvent strict waste disposal laws within the United States by shipping wastes to poor countries willing to accept toxic cargoes in exchange for much-needed cash? Or is this an ethically indefensible action that imperils America's image as a nation in good standing with the international community? Thus far, federal policy on this issue has wavered between a weak regulatory approach and an approach placing greater emphasis on the marketplace as the preferred locus of resource decisions.[38]

During the Carter Administration, EPA's international office took the position that foreign policy decisions with an environmental component should be consistent with the human rights theme articulated by the president. Under this doctrine, administrators called for restrictions or in some cases an outright ban to be placed on the exportation of hazardous wastes to countries ill equipped to manage them efficiently or safely. EPA essentially won the policy contest against the Departments of State, Commerce, and Agriculture, which pushed for the principle of nonintervention in international markets.[39]

The latter course of action was subsequently embraced by the Reagan Administration and has remained largely intact to the present time.[40] An RCRA amendment adopted in 1980 allowed industry to export wastes to other nations provided that EPA was given prior notification "about planned shipments, the type and quantity of waste, and...plans for treatment, storage and disposal in the receiving country."[41] But the benefits of this requirement were largely restricted to information gathering. There was no effort on the part of federal administrators to use this provision as an early warning system to communicate the possibility of harm to humans or environmental quality.

A potentially useful policy corrective was added through the Hazardous and Solid Waste Amendments of 1984 (HSWA) and regulations promulgated in 1986. HSWA called for the prohibition of waste exports from the United States without the written consent of relevant public officials within the recipient countries. Another change allowed the federal government to enter into bilateral agreements pertaining to the exchange of wastes. As a result, agreements were successfully negotiated with both Canada and Mexico in late 1986.

However, a closer examination of how policy is actually implemented indicates that economic rather than environmental protection goals have taken precedence. For example, an evaluation of EPA program activities has revealed widespread noncompliance with agency regulations, the failure to report objections raised by importing countries to exporter firms, and the neglect of data-gathering responsibilities.[42] Second, the volume of waste exports to Canada has risen substantially with minimal controversy, a result that has permitted U.S. companies to avoid increasingly large off site disposal costs within this country while enriching the coffers of waste management firms in Ontario and Quebec. Finally, agency administrators have not been given a green light to inform waste-importing nations about the environmental risks posed by forthcoming shipments of toxic cargo.[43]

Congress

The implementation of hazardous waste policies by EPA also can be affected by the agency's relationship with Congress. I have already alluded to one of the more critical aspects of legislative involvement—policy design. The assignment of program management authority; the degree of latitude extended to EPA on the development of regulations; the hierarchical integration of federal, state, and local agencies; and the expansion or restriction of policy actors via the provision of public participation represent dimensions of activity that are embedded within the language of hazardous waste statutes.

Another source of congressional involvement in EPA decision making is the allocation of staff and budgetary resources (to be discussed more fully in the following section). Several writers have commented on the growing gap between programmatic responsibilities assigned to EPA by Congress and the availability of resources to carry out these programs. This complication is reinforced further by the additional staff, time, and money diverted to the preparation of Regulatory Impact Assessments (RIAs) for proposed regulations.

A third area of legislative impact involves the occasional use of oversight hearings. High-visibility incidents such as the discovery of dioxin-contaminated roads in Times Beach, Missouri, and the political manipulation of Superfund contracts by EPA Administrator Burford and Assistant Administrator Lavelle resulted in a spate of hearings designed to identify problems related to the administration of hazardous waste programs. Four oversight hearings were held in October 1981 in both chambers "to determine whether the Reagan Administration is using deep budget cuts and regulatory 'reform' to retreat from its congressionally mandated pollution control duties."[44]

Fourth, the strategic use of information and policy analysis can contribute to congressional influence over agency decisions.[45] Testimony is often obtained in the form of reports and analyses from an array of interested parties at oversight hearings. Other types of information are gleaned from studies conducted by agencies within the legislative branch—notably the Office of Technology Assessment (OTA), the Congressional Research Service (CRS), the Congressional Budget Office

(CBO), and the General Accounting Office (GAO). OTA has authored several comprehensive analyses of hazardous waste policy in general and Superfund in particular, while the other organizations often focus on narrower administrative problems associated with program management decisions.

Judiciary

Yet another source of institutional impact on EPA is presented by the federal courts. Judicial review of EPA actions has affected the range of administrative choices in several ways. Perhaps the most commonly mentioned factor is the expression of the courts' role in reviewing the substance of regulations and other administrative decisions.[46] This review involves a closer look at the facts of the case to ensure that agencies have considered most or all relevant factors and have, in some instances, provided sufficient documentation to justify courses of action.[47]

Substantive review requires a degree of evaluative judgment on the part of judges. This is particularly evident in federal court decisions involving Superfund.[48] The statute is silent on a rather significant point: the apportionment of financial liability among multiple parties responsible for hazardous waste disposal at an abandoned dump site. However, this question was addressed in *U.S. v. Chem-Dyne Corp.*, an important case that gave policy direction to EPA administrators by inferring congressional intent on the basis of prior statements written by the original legislative sponsors of the statute.[49]

A key consequence of this ruling was that financial burden sharing must be allocated under the basis of the joint and several liability doctrine, an approach that places the burden of proof on the defendants. Officials representing waste-generating firms argued that Congress would not have permitted a potentially costly topic of this sort to be inferred. Intent, they contended, would have been incorporated within the body of the statute itself. Nevertheless, subsequent court decisions have affirmed the legal validity of EPA's use of this doctrine.

Other examples of federal court impacts on the administration of hazardous waste programs can be attributed to the occasional willingness of judges to question the scientific expertise of EPA staff. In the words of O'Leary:

> In one toxics case...the judge ordered the plaintiff environmental group to enforce the court order by commencing a new action should the EPA not comply with the court mandate. To facilitate the filing of a new lawsuit against the EPA, the judge waived a congressionally mandated sixty days' notice provision, in direct violation of the statute. In this instance, the

judge coupled detailed judicial supervision of the agency through an ongoing affirmative decree with supervision by the plaintiffs.[50]

Third, court decisions have influenced the establishment of programmatic priorities within EPA. Compliance with court orders takes precedence over staff policy preferences within various bureaus. As a result, agency actions are more likely to reflect sensitivity to judicial decrees than an overall assessment of which problems require immediate attention on the basis of risk or congressionally mandated deadlines.[51] Moreover, this problem is reinforced by the failure of Congress to allocate funds or staff for the purpose of meeting court mandates. This deprivation stems from a combination of general fiscal austerity plus the difficulty of integrating supplemental budget requests within existing EPA planning and budgetary cycles.[52]

Interest Groups and the Media

Nongovernmental political actors such as interest groups and the media also can influence program decision making. Organizations vary considerably in terms of the breadth of their policy concerns and the types of strategies they use to influence administrative decisions. Multipurpose environmental groups such as the Sierra Club, the Audubon Society, the Environmental Defense Fund (EDF), the Natural Resources Defense Council (NRDC), and the National Wildlife Federation (NWF) deal with a large number of issues ranging from resource conservation to environmental protection and possess the resources to use a variety of political tactics as well.

Also active on hazardous waste issues are organizations favoring a middle-ground position, such as the Hazardous Waste Treatment Council and the Association of State and Territorial Solid Waste Management Officials. Their representatives often push for solutions that balance public or ecological health with economic feasibility. Other environmental organizations (such as the Keystone Center or the Conservation Foundation) also focus on less adversarial approaches to achieve environmental goals—namely, the use of mediation between contending parties or economic incentives to bring about industry compliance.

One of the more significant forms of organizational activity revolves around efforts to shape the direction and content of rules proposed by EPA. Environmental organizations actively lobby EPA during the earlier phases of rulemaking in an effort to force regulated industries to bear more of the costs associated with the abatement and cleanup of hazardous wastes. Likewise, trade groups such as the Chemical Manufacturers' Association and the American Petroleum

Institute, as well as specific industry spokespeople will vigorously oppose the promulgation of regulations that may divert resources away from research and development, production, or marketing activities to meet mandated environmental quality goals.

A case in point is the EPA's proposed rule governing the incineration of hazardous waste in cement kilns. A number of cement companies have sought permission to reduce their fuel costs in the production phase by mixing hazardous wastes with more conventional fuels. Industry officials contend that a net gain in both economic efficiency and environmental quality is achieved through this option, because incineration promotes recycling of wastes that would otherwise be disposed of in land-based tsd facilities. They also claim that more of the pollutants are disposed of cleanly, which in turn reduces the likelihood of land or groundwater contamination.

Environmentalists argue that burning waste-laden fuels does not actually solve pollution problems but reorients them from one medium (land) to another (air). Moreover, pollutants are not likely to be fully destroyed because the kilns are not engineered with hazardous waste disposal as a primary objective; further, an additional byproduct from this process, fly ash, requires removal to a permitted facility. The intensity of the rulemaking debate over incineration has been exacerbated by a SARA-based requirement that each state develop sufficient waste-disposal capacity to manage toxic wastes generated within its borders or forfeit the opportunity to receive federal funding for Superfund cleanup activities.

While much group activity is invested in rulemaking processes, both lawsuits and testimony at congressional hearings are commonly used tactics aimed at influencing the implementation of hazardous waste programs. Litigation is clearly a preferred strategy for many organizations. EPA officials are often whipsawed between lawsuits filed by industry officials seeking to avoid or delay regulatory compliance costs and those initiated by environmental groups to ensure that compliance deadlines are not ignored by agency administrators. A number of groups also offer testimony at oversight hearings conducted by Congress to track EPA's administrative activities.

Other approaches include initiating research projects dealing with policy or management issues that will be made available to EPA officials and key legislators and assuming the role of watchdog in monitoring the actions of EPA and other federal agencies. The latter strategy requires the aid of media sources to publicize the seriousness of environmental contamination problems stemming from the nonenforcement of hazardous waste laws. If negative publicity of this sort does not spur the desired administrative response from EPA or the state agency in charge of hazardous waste programs, then litigation can be pursued as a fallback strategy.

RESOURCE CONTEXT

Programs require resources; that is, money and people. EPA currently spends $1.8 billion on hazardous waste programs involving approximately 5,000 work-years.[53] But the importance of resource levels in the decision-making process can be more readily appreciated if expenditures are examined in relation to mandated responsibilities. In other words, how much is enough in view of political pressures, technological uncertainty, and statutory deadlines?

On the one hand, it is important to acknowledge the point that money alone cannot ensure that administrative tasks will be completed expeditiously. Delay in RCRA and Superfund decision making has occurred because of scientific disagreement over program objectives as well as the questionable use of deadlines to "force" agency action without a clear understanding of how long it takes in the form of person-years to accomplish a given assignment.[54] For example, Dower cautions against proceeding too quickly on Superfund site cleanups, because too little is known about the risks associated with individual sites and, as a result, the most appropriate method of resolving the problem.[55]

On the other hand, it is argued that Congress has tended to assign new regulatory responsibilities to EPA without a commensurate increase in resources. This was especially pronounced during the 1980s when the Reagan Administration sought—and received—deep cuts in expenditures for pollution control programs administered by EPA. According to Kraft and Vig, federal spending within this category declined by a third between 1980 and 1990.[56] They also reveal that "in constant 1982 dollars, EPA's operating budget in fiscal 1990...was scarcely any higher than it had been in fiscal 1975, despite the important new responsibilities given to the agency by Congress."[57]

As we look more closely at hazardous waste budgets over time, a number of points bear mention. First, the differences between the last Carter budget and the initial Reagan budget are striking in terms of dollar amounts and programmatic priorities. By the end of fiscal 1981, the federal government spent $141 million on RCRA, while Reagan's RCRA budget for fiscal 1982 was less than $120 million—a 23-percent cut. This budgetary decrease was coupled with a governmentwide hiring freeze in domestic programs, which inhibited efforts to recruit or replace critical personnel.[58]

Second, greater emphasis was placed on some programmatic activities rather than others. Especially hard hit during the early years of the Reagan Administration was the enforcement budget, a direct consequence of the president's belief that the federal government should pursue a less adversarial approach vis-à-vis the private sector in the implementation of environmental protection policies. Conflict was considered counterproductive, as the administration clearly empathized with industry concerns about regulatory compliance costs. Moreover, EPA Administrator Burford and Assistant Administrator Lavelle were convinced that affected firms would respond to a more flexible governmental policy by acting responsibly in their management of hazardous wastes.[59] This view was reflected in an 86-percent decrease in funding for enforcement actions between 1981 and 1983.[60]

Another source of budgetary cutbacks that was somewhat controversial in the early 1980s was research and development expenditures. The 1981 dollar amount for this activity ($28 million) was reduced by nearly a quarter from fiscal 1982 through fiscal 1984. This strategy has been described in terms akin to penny wise–pound foolish by critics who emphasized the need for strengthening the scientific foundation of environmental regulations. If regulatory reform was truly a goal of Reagan Administration officials, then research and development funding for standard setting should have been expanded rather than reduced.[61] In addition, similar areas of concern, such as the budget for independent research, were also cut substantially, leading one analyst to conclude that Burford's primary goal was less a commitment to better science than an intent to implement the Reagan/Stockman policy of deregulating domestic social programs.[62]

Funding and staffing for hazardous waste programs fared somewhat better under Burford's successors, William Ruckelshaus and Lee Thomas. Between 1984 and 1988, budget outlays for abatement, control, and compliance and for research and development increased modestly, while enforcement remained a lower-priority activity. In part, increased funding reflected expanded responsibilities for EPA mandated under the reauthorization of RCRA in 1984 and Superfund in 1986.

Since the election of President George Bush in 1988, the resource base for hazardous waste programs has remained steady in some areas but has improved in others. Funding for research and development within the RCRA program actually declined slightly, reflecting the argument that much of the research needed to support regulatory development has been completed. On the other hand, the Bush Administration has pushed for sizable increases in spending for the Superfund program to accelerate the implementation of emergency response actions and permanent cleanups. Another item given budgetary priority by EPA Administrator William Reilly has been enforcement. Increased outlays were received not only for the expansion of enforcement actions within the RCRA and Superfund programs but within the environmental protection division of the Department of Justice as well.[63]

How much is enough in terms of resources remains a bone of contention between those who argue that current funding levels reflect a reasonable effort to meet program needs and those who insist that the gap between EPA's mandated responsibilities and the number of person-years (i.e., staff) required to carry them out is still too large. Another concern that has emerged recently is a lack of continuity in HSWA program operations because of staff turnover. Professional employees with experience in regulatory decision making have become increasingly frustrated with the dearth of career advancement options, few training opportunities, low pay, and the lack of glamour attached to hazardous waste regulation in comparison to more visible pollution-control issues. As a result, a sizable number of skilled professionals have obtained internal transfers from RCRA to the Superfund program.[64]

CONCLUSIONS

A discussion of EPA's experience with an array of political actors and institutions with a stake in hazardous waste policy decisions provides us with an excellent example of Pressman and Wildavsky's admonition about the link between the "complexity of joint action" and subsequent administrative problems.[65] It is somewhat paradoxical that EPA officials were given a broad grant of decision-making authority by Congress in view of the need for experts to resolve a number of technically difficult questions only to be hamstrung by the necessity of obtaining clearance or approval for their decisions from a host of political players. This is reflected in the amount of delay associated with the promulgation of regulations and the issuance of permits early on. Responsiveness has been pursued at the expense of efficiency.

Which factors stand out in our efforts to account for administrative decision making within EPA? Perhaps the most prominent—and obvious—precipitator of policy direction and behavior is the appointment of the administrator and the other top agency officials. Marked fluctuations in agency performance accompanied the changeover from Douglas Costle to Anne Burford and William Ruckelshaus. In addition, the administrator is in a position to influence other institutions, actors, and decisions affecting what EPA does, such as the nature and frequency of communications with regional staff, state officials, and the regulated community; budgetary recommendations; the maintenance of a good working relationship with key political institutions and constituencies; the scope and direction of regulations; and the emphasis given to enforcement.

Major sources of external influence on EPA decision making include Congress, the OMB, and the federal courts. Congressional guidance is most keenly felt in the design of hazardous waste statutes, especially the reauthorized versions of RCRA and Superfund, and in the allocation of budgetary and staff resources. From an EPA perspective, the effects of legislative intervention are mixed. On the one hand, there was some acknowledgment by federal legislators that the pace of implementation had been adversely affected by a lack of resources. Thus, HSWA has received a modest but steady increase in appropriations since the mid-1980s, while Superfund managed to obtain a fivefold increase in funding.

On the other hand, Congress has not given EPA the tools needed to induce other federal agencies to comply with hazardous waste laws. Insufficient legal authority at this level also undercuts enforcement efforts in state capitols, since policymakers are more than willing to suggest a "glass houses" analogy. Aggressive regulation should not be undertaken by EPA officials in City X until they can demonstrate that federal agencies have adopted good environmental management practices. Agency officials also lament the loss of considerable administrative flexibility with the adoption of HSWA and SARA. Congress vented its frustration with EPA's lack of progress in handling hazardous waste problems by relying more strongly on the use of goals and timetables as an action-forcing strategy.

Constraints on EPA's programmatic choices at both ends of the decision-making spectrum are also presented by the OMB and the federal courts. The regulatory agenda proposed by agency officials must now be reviewed by OIRA before the information-gathering process can begin. Action agendas also are influenced by federal court decisions over challenges to regulations initiated by industry or environmental groups. Even if the controversy does not address issues considered to be of greater priority to EPA officials, resources are devoted to satisfying judicial edicts first. Proposed rules also can be altered to meet the concerns of OMB officials.

NOTES

1. This point is argued effectively in Bruce Ackerman and William T. Hassler, *Clean Coal/Dirty Air* (New Haven, Conn.: Yale University Press, 1981).

2. Charles Davis, "Approaches to the Regulation of Hazardous Wastes," *Environmental Law, 18*, No. 3 (1988), pp. 511–12.

3. A useful discussion of this approach can be found in David M. Welborn, "Conjoint Federalism and Environmental Regulation in the United States," *Publius, 18* (Winter 1988), pp. 27–43.

4. Samuel Epstein, Lester Brown, and Carl Pope, *Hazardous Waste in America* (San Francisco: Sierra Club Books, 1982).

5. Consult Water Rosenbaum, "The Politics of Public Participation in Hazardous Waste Management." In James P. Lester and Ann O'M. Bowman, eds., *The Politics of Hazardous Waste Management* (Durham, N.C.: Duke University Press, 1983), pp. 176–95.

6. N.a., "Congress Tightens Hazardous Waste Controls," *1984 CQ Almanac* (Washington, D.C.: Congressional Quarterly, 1985), p. 307.

7. Alfred Marcus, *Promise and Performance: Choosing and Implementing an Environmental Policy* (Westport, Conn.: Greenwood Press, 1980), Chapter 3.

8. Ackerman and Hassler, *Clean Coal / Dirty Air*, Chapter 1.

9. Richard W. Waterman, "Reagan and the EPA: Revolution and Counter–Revolution." 1987 Midwest Political Science Association Conference Paper, Chicago.

10. Marcus, *Promise and Performance*, Chapter 4.

11. Ibid. Chapter 4.

12. Ibid. Chapter 4.

13. U.S. Government Manual (Washington D.C.: Government Printing Office, 1990), pp. 554–55.

14. This point is discussed in Patricia Crotty, "Assessing the Role of Federal Administrative Regions: An Exploratory Analysis," *Public Administration Review, 48* (March/April 1988), pp. 642–48.

15. J. Clarence Davies, "Environmental Institutions and the Reagan Administration." In Norman Vig and Michael Kraft, eds., *Environmental Policy in the 1980s* (Washington, D.C.: CQ Press, 1984), pp. 143–60.

16. Norman J. Vig, "Presidential Leadership: From the Reagan to the Bush Administration." In Norman Vig and Michael Kraft, eds., *Environmental Policy in the 1990s* (Washington, D.C.: CQ Press, 1990), pp. 33–58.

17. Ibid.

18. Steven Cohen, "Defusing the Toxic Time Bomb: Federal Hazardous Waste Programs." In Norman Vig and Michael Kraft, eds., *Environmental Policy in the 1980s* (Washington, D.C.: CQ Press, 1984), p. 282.

19. Ibid

20. Ann O'M. Bowman, "Superfund Implementation: Five Years and How Many Cleanups?" In Charles E. Davis and James P. Lester, eds., *Dimensions of Hazardous Waste Politics and Policy* (Westport, Conn.: Greenwood Press, 1988), p. 140.

21. This problem is discussed in Seymour Schwartz, Wendy Cockovich, Cecelia Fox, and Nancy Ostrom, "Improving Compliance with Hazardous Waste Regulations among Small Businesses," *Hazardous Waste and Waste Materials, 6* (Summer 1989), pp. 281–96.

22. Edward Fuchs, *Presidents, Management and Regulation* (Englewood Cliffs, N.J.: Prentice Hall, 1988), Chapter 4.

23. Subcommittee on Oversight and Investigations, House Committee on Energy and Commerce, *OMB Review of EPA Regulations*, 99th Congress, 2nd Session, 1986.

24. Ibid.

25. Ibid., p. 24.

26. This point is discussed in James O'Reilly and Phyllis Brown, "In Search of Excellence: The Future of OMB Oversight of Rules," *Administrative Law Review* (Fall 1987).

27. Ibid.

28. Subcommittee on Oversight and Investigations, *OMB Review of EPA Regulations*, p. 26.

29. Ibid.

30. Ibid.

31. General Accounting Office, *Environmental Funding: DOE Needs to Better Identify Funds for Hazardous Waste Compliance* (GAO/RCED-88-62, December 1987), p. 5.

32. General Accounting Office, *Hazardous Waste: Federal Civil Agencies Slow to Comply with Regulatory Requirements* (GAO/RCED-86-76, May 1986), p. 24.

33. General Accounting Office, *Efforts to Clean Up DOD-Owned Inactive Hazardous Waste Sites* (GAO/NSIAD-85-41, April 12, 1985).

34. Ibid., Chapter 4. Perhaps the best analysis of problems with military and civilian agency management of hazardous wastes is contained in Congressional Budget Office, *Federal Liabilities Under Hazardous Waste Laws* (Washington, D.C.: Government Printing Office, May 1990).

35. Karen B. Wiley and Steven L. Rhodes, "Decontaminating Federal Facilities: The Case of the Rocky Mountain Arsenal," *Environment, 29* (April 1987).

36. Ibid., p. 29.

37. See, e.g., Charles Davis and Joe Hagan, "Exporting Hazardous Wastes: Issues and Policy Implications," *International Journal of Public Administration* (Fall 1986); and Henry Shue, "Exporting Hazards," *Ethics* (July 1981).

38. The importance of market forces in shaping the relationship between chemical firms and governmental authorities in Western countries, including the United States, is discussed in Thomas Ilgen, "Better Living through Chemistry: The Chemical Industry in the World Economy," *International Organization 37* (Autumn 1983).

39. Davis and Hagan, "The Export of Hazardous Wastes."

40. Ibid.

41. A market-oriented approach toward hazardous waste exports has been treated as an economic safety valve by industry officials plagued with spiraling waste-disposal costs in the United States. The 1984 amendments to RCRA significantly restricted the use of both deep-well injections and landfills as waste disposal options. As a result, the average cost per ton for toxic waste disposal rose from $15 in 1980 to more than $200 in 1987. While the intent of these changes was to promote the use of source reduction as a management strategy, an unintended consequence has been a sharp rise in export applications. Notifications of industry intent to ship wastes overseas received by EPA increased from 12 in 1980 to 522 in the first six months of 1988. See Mary Deery Uva and Jane Bloom, "Exporting Pollution: The International Waste Trade," *Environment 31* (June 1989), especially pp. 4–5.

42. Barry G. Rabe, "Exporting Hazardous Waste in North America," *International Environmental Affairs 3* (Spring 1991), p. 114.

43. Ibid.

44. Cited in Mary Etta Cook and Roger Davidson, "Deferral Politics: Congressional Decision-Making Issues in the 1980s." In Helen Ingram and Kenneth Godwin, eds., *Public Policy and the Natural Environment* (New York: JAI Press, 1985), p. 7l.

45. The tension arising between EPA and the federal courts, as well as a number of issues arising from differing views of what the appropriate role of the federal courts should be in the review of administrative decisions, is analyzed in Shep Melnick, *Regulation and the Courts: The Case of the Clean Air Act* (Washington, D.C.: Brookings Institute, 1983).

46. See, for example, Donald Horowitz, *The Courts and Social Policy* (Washington, D.C.: Brookings Institute, 1977); and Martin Shapiro, "On Predicting the Future of Administrative Law," *Regulation, 6* (May/June 1982).

47. The need for documentation to justify high-level administrative decisions was established in a Supreme Court decision, *Citizens to Preserve Overton Park v. Volpe*, 401 U.S. 402 (1971).

48. Werner F. Grunbaum, "Developing a Uniform Federal Common Law for Hazardous Waste Liability," *Policy Studies Journal 14* (September 1985), pp. 132–39.

49. 572 F. Supp. 802.

50. Rosemary O'Leary, "The Impact of Federal Court Decisions on the Policies and Administration of the U.S. EPA," *Administrative Law Review, 41* (Fall 1989), p. 561.

51. Ibid. See also Melnick, *Regulation and the Courts.*

52. Ibid., pp. 563–64.

53. These figures are based on EPA's hazardous waste and Superfund outlay estimates for 1990, as reported in Environmental Protection Agency, *Summary of the 1991 Budget* (January 1990).

54. Rosenbaum, *Environmental Politics and Policy*, Chapter 5.

55. Roger C. Dower, "Hazardous Wastes." In Paul R. Portney, ed., *Public Policies for Environmental Protection* (Washington, D.C.: Resources for the Future, 1990), pp. 151–94.

56. Michael Kraft and Norman Vig, "Environmental Policy from the Seventies to the Nineties: Continuity and Change." In Norman Vig and Michael Kraft, eds., *Environmental Policy in the 1990s* (Washington, D.C.: CQ Press, 1990), p. 19.

57. Ibid.

58. Cohen, "Defusing the Toxic Time Bomb," p. 279.

59. Ibid., p. 285.

60. U.S. Office of Technology Assessment, *Technologies and Management Strategies for Hazardous Waste Control* (Washington, D.C.: Government Printing Office, 1983), p. 340.

61. Robert V. Bartlett, "The Budgetary Process and Environmental Policy." In Norman Vig and Michael Kraft, eds., *Environmental Policy in the 1980s* (Washington, D.C.: CQ Press, 1984), pp. 132–36.

62. Richard N.L. Andrews, "Deregulation: The Failure at EPA." In Norman Vig and Michael Kraft, eds., *Environmental Policy in the 1980s* (Washington, D.C.: CQ Press, 1984), p. 176.

63. EPA, *Summary of the 1990 Budget.*

64. U.S. Environmental Protection Agency, *The Nation's Hazardous Waste Management Program at a Crossroads: The RCRA Implementation Study* (Washington, D.C.: Government Printing Office, July 1990), pp. 91–92.

65. Jeffrey Pressman and Aaron Wildavsky, *Implementation*, 3rd ed. (Berkeley: University of California Press, 1984).

EPA AND THE IMPLEMENTATION OF HAZARDOUS WASTE POLICIES

We now turn to the role played by EPA in shaping the implementation of RCRA and Superfund. Although administrative decisions are influenced by a myriad of political actors and institutions, there are nonetheless a number of levers that can be pulled by agency officials. First, the manner in which program objectives are communicated to regional EPA administrators and state officials is important because it sets the tone for subsequent decisions. A second area of responsibility involves the delegation of hazardous waste programs to state environmental protection agencies. Third, EPA administrators initiate rules designed to aid in the process of reducing general policy prescriptions to guidelines for field-level decisions. Last but not least is the monitoring and enforcement function, which is a collaborative exercise involving EPA, state agencies, the U.S. Department of Justice, and the courts, among others.

COMMUNICATION OF PROGRAM OBJECTIVES

A key responsibility of any cabinet secretary or EPA administrator is the communication of policy goals to others throughout the organization. The enforcement of regulatory policies is particularly dependent upon the ability of top administrators to convey the message that noncompliance on the part of regulated parties will not be tolerated. Establishing regulatory credibility up front during the initial phase of a new presidency increases the likelihood of voluntary compliance and sensitizes state administrators to federal program priorities.

A frequently cited example is the decision by former EPA Administrator William Ruckelshaus to initiate a few highly publicized lawsuits against steel companies found to be in violation of the Clean Air Act in the early 1970s.[1] In part, the early use of litigation as an enforcement strategy was intended as a form of shock therapy. EPA, as a newly formed regulatory agency, would be better positioned to achieve its mission if its dicta were taken to heart by industry officials. Another hoped-for political byproduct of this strategy was the added measure of influence enjoyed by Ruckelshaus in subsequent negotiations with individuals representing pollution-generating firms.[2]

The importance of communication as a means of enforcement can be described more easily by focusing on specific aspects of the decision-making process.[3] Message content involves a broader view of policy than the information gleaned from statutory language. Would-be receivers, notably industry and state government officials, are attuned to an array of cues including policy-based information, organizational change at EPA, resource-allocation decisions, and administrative relationships involving key personnel inside the agency and administrators or elected officials from other institutions.[4]

The Burford-Lavelle era in the early 1980s provides an example of problems caused by sending multiple signals. From the outset, the EPA was in a state of organizational disarray. Burford initially chose to abolish the Office of Enforcement before restoring it within the agency's legal office, a signal that voluntary compliance was the favored means of handling waste-disposal problems. She and other appointees also chose to rely less on advice from Congress or careerists within EPA, preferring to solicit information from Reagan Administration loyalists or from industry representatives. When these incidents are coupled with decisions such as budget cuts for hazardous waste program operations and the suggested elimination of grants to state agencies, it is difficult to avoid the conclusion that a clear message was sent—intended or not—that organizational efforts to detect and punish industry violators of hazardous waste laws were less important than program deregulation and decentralization.

The form of the message and the credibility of the public officials in charge can also affect the sense of urgency attached to enforcement actions by industry and state officials. William Ruckelshaus was a popular choice to succeed Burford because of the widespread belief that he had effectively administered pollution-control laws during his earlier stint as EPA administrator. Initial actions taken by Ruckelshaus closely resembled the administrative style he exhibited in the early 1970s. Notice was given that EPA's role as the primary enforcement agency for environmental protection laws would be restored.

The renewal of enforcement as a central policy goal was revealed in a rather colorful speech to state officials in which Ruckelshaus explained the agency's responsibilities in terms akin to the good cop–bad cop metaphor. EPA could be usefully portrayed by state regulators as a "gorilla in the closet" that could be unleashed if polluting firms refused to comply with environmental policies.[5] Moreover, symbolic gestures of this sort were reinforced by the development of a new management system for tracking federal and state enforcement actions and an increase in the number of lawsuits filed against corporate violators.

Ruckelshaus was also cognizant of the need to restore and communicate to others a picture of EPA as a committed, competent, and effective agency. He persuaded a number of experienced administrators such as Deputy Director Alvin Alm to come on board, thereby infusing the agency with doses of energy and expertise. Efforts to cultivate a closer relationship with Congress and the media also were begun. In short, the range and consistency of actions taken by Ruckelshaus produced the desired effects of bolstering the morale of personnel within EPA and communicating to the regulated community his intention to enforce the law. Deterrence replaced assumptions of good faith as the preferred means of inducing compliance with hazardous waste programs. This style of communication was maintained by Lee Thomas, who succeeded Ruckelshaus as EPA administrator in 1985, and by Bush appointee William Reilly.[6]

DELEGATING MANAGEMENT AUTHORITY TO THE STATES

Federal law requires a substantial state role in the administration of hazardous waste policies. This is particularly true for RCRA, one of several federal policies with a partial preemption policy design.[7] After developing uniform environmental quality standards, EPA is authorized to review state applications for the assumption of program management responsibilities. If EPA officials are persuaded that a given state can effectively administer the law, the decision to delegate regulatory authority is made. EPA officials are still obliged to oversee state enforcement efforts and to provide backup if needed.

Failure on the part of state officials to carry out their part of the intergovernmental bargain may result in the preemption of management authority by EPA. Under these circumstances, EPA would deploy administrators from the appropriate regional office to implement the state program.

Let us turn to a more specific examination of the process by which EPA delegates RCRA program management authority to the states. If state officials apply for authorization, the package must demonstrate that the proposed program is equivalent to and no less stringent than the federal program and that the law will be effectively enforced. The effect of this preclearance requirement is to keep a considerable degree of administrative control in the hands of federal officials.[8]

It appears that the inclination of federal officials to entrust state administrators with significant program management authority has varied across and within presidential administrations. EPA officials within the Carter Administration were slow to transfer decision-making control to the states, partly because of the newness and complexity of hazardous waste problems and partly because of a skeptical attitude toward the ability and willingness of state officials to enforce environmental regulations.[9] Relatively few states achieved full authorization to manage RCRA (those that did will be referred to hereafter as *primacy states*).

A major change in the nature and frequency of delegation decisions occurred when Anne Burford became EPA administrator. Her philosophical leanings toward greater state involvement were quite compatible with President Reagan's "New Federalism" focus, and she expressed an intention to "restore the states to their rightful place as partners with the federal government in policymaking as well as policy implementation."[10] The immediate transfer of environmental decision-making authority to state-level officials became a priority concern within EPA, and decision rules were altered to accelerate the process. By mid-1983, 37 states had received EPA approval to administer RCRA. All but ten of these were authorized during Burford's tenure.[11] State officials were clearly pleased with the increase in programmatic responsibility but were considerably less enthusiastic about the administration's proposal to reduce and eventually eliminate RCRA grants.

In 1984, the delegation process began to decelerate for two reasons. First, William Ruckelshaus's goal as the newly appointed EPA administrator was to bring about greater balance between the goals of enforcement and decentralization. Efforts to achieve a clearer delineation of jurisdictional responsibilities were undertaken. A federal task force recommended that the role of EPA be restricted to research, standard setting, assistance, and oversight, with the states taking charge of day-to-day administrative activities.[12]

A second factor helping to account for the slower transfer of authority was the adoption of the Hazardous and Solid Waste Amendments of 1984 (HSWA), which incorporated changes in the authorization process. Prior to the enactment of HSWA, states receiving EPA approval to manage RCRA were given considerable latitude to administer the entire program subject to federal oversight. However, recently adopted federal regulations did not become effective in these states until equivalent regulations were promulgated by the agency in charge of hazardous waste management.[13] Failure on the part of state officials to maintain consistency between their own regulatory requirements and those emanating from Washington within a given period of time could result in preemption of state management authority by EPA.

On the other hand, one of the objectives of HSWA was to ensure that protecting the public from exposure to hazardous chemicals would not be unduly compromised by interstate variation in regulatory lag; that is, the amount of time needed by state officials to respond to changing federal requirements. Under HSWA, newly promulgated federal regulations became effective immediately, in primacy as well as nonprimacy states. This change addressed the problem of regulatory lag by recentralizing key aspects of enforcement decisions. Until states managed to develop equivalent regulations, a dual system of enforcement involving both EPA and state administrators would be in place.[14]

In short, an examination of RCRA over time suggests that public officials within the executive branch and in Congress are willing, and in some cases enthusiastic, proponents of delegating hazardous waste program responsibility to the states, but views pertaining to "how much" and "how quickly" vary across and within presidential administrations. A more restrictive approach, which allows the states to enforce hazardous waste programs only after meeting regulatory requirements established by Congress and EPA, has been characteristic of the Carter Administration and of EPA administrators William Ruckelshaus, Lee Thomas, and William Reilly.

Despite the "partnership" between the federal government and the states forged by the architects of partial preemption, EPA officials have maintained control over much of the RCRA program by promulgating highly detailed, prescriptive regulations that set minimum program standards for states. This is reinforced by EPA's retention of sufficient legal authority to review and/or veto most state regulatory decisions. While the Reagan Administration's commitment to state-centered solutions for regulatory problems was initially perceived as a positive sign by state officials hoping for a more equal role in the intergovernmental distribution of programmatic responsibilities, it is unclear whether an actual transfer of decision-making power took place. A study of state administrators in the Great Lakes region indicated that most did not believe that EPA was willing to give any weight to hazardous waste policy issues of concern to state officials.[15]

Unlike the RCRA or HSWA statutes, which contain a partial preemption format, Superfund is not constructed in ways that permit the a priori designation of state or federal responsibility. Either jurisdiction can take the lead in emergency response or permanent cleanup projects. Because of high political visibility and the economic importance of cost allocation decisions, NPL site management decisions often attract numerous participants from EPA and state government as well as individuals representing industry and environmental organizations. According to Bowman, EPA is most likely to act alone on emergency response actions, while permanent cleanup decisions involve negotiations between EPA, state administrators, and others.[16]

REGULATORY ACTIONS

Rulemaking

Rulemaking, like the delegation of program management authority, is based on the specification of conditions that must be achieved before allowing state officials to assume control. The development of regulations by EPA officials is designed to provide the necessary "connective tissue" between the broad policy goals embedded in RCRA and Superfund and the day-to-day administrative tasks carried out by field representatives. More specifically, rules offer essential guidance on questions dealing with the range and type of wastes to be regulated or exempted; criteria for the management of industrial wastes on facility grounds; safety measures to be incorporated in the transport of wastes off-site; standards for the construction, operation, and closure of tsd (treatment, storage, and disposal) facilities, insurance or other financial requirements for tsd facilities; listing and delisting procedures for NPL sites and the appropriate avenues for public involvement in the decision-making process, among others.

Historically, EPA has used a rulemaking process that is detailed, arduous, and cumbersome (see Table 4–1). It begins with a draft authorized by administrators within the relevant program office. The document contains the objective of the proposed rule, the options to be considered, plans for coordinating the rule within EPA, public involvement efforts, and a schedule. Phase II consists of forming a working group within the lead office and developing the analyses in support of the regulation (including a regulatory impact assessment for rules expected to produce compliance costs of $100 million or more).

Input is then solicited from an array of players, including other executive branch agencies; organizations representing industries, consumers and the environment; state and local government associations; Congress; the OMB; and occasionally science advisory panels. Next, a

TABLE 4–1 The Rulemaking Process in EPA

Phase I: Start-up (one month)
Initial decision to begin a rulemaking proceeding
Organization of working groups
Preparation of background and preliminary materials
Creation of public docket

Phase II: Development (six months)
Preparation and review of development plan
Consultation with EPA management and technical advisory groups
Consultation with external contractors and consultants
Consultation with interest groups, federal agencies, congressional committees
Internal management review: Steering Committee
red border
administrator
Publication of advanced notice of proposed rulemaking in Federal Register

Phase III: Preparation and review of proposed rulemaking package (ten months)
Working group preparation of proposed rule and supporting materials
Internal management review: Steering Committee
red border
administrator
OMB review of paperwork requirements
OMB review of proposed rulemaking package and regulatory analyses
Publication of notice of proposed rulemaking in Federal Register
Public comment period

Phase IV: Preparation and review of final rulemaking package (20 months)
Working group preparation of final rule and supporting materials
Internal management and OMB reviews—same as for phase III
Publication of final rule in Federal Register
Legislative and judicial review

Source: Gary C. Bryner, *Bureaucratic Discretion* (New York: Pergamon Press, 1987), p. 99.

rather intensive internal management review occurs; it involves the EPA administrator, assistant and regional administrators, agency attorneys, and the office directors. At this point, advance notice of the proposed rule is published in the Federal Register.

Once the regulatory package passes muster internally, it is submitted to OMB for review. OMB has 60 days to approve the proposed rule or to recommend changes. Copies of the proposed rules are occasionally sent to House and Senate committees with jurisdiction over hazardous waste issues as well. If the regulation survives the political gauntlet through this stage, it is signed by the administrator and published in the Federal Register. A public comment period lasting between 60 and 130 days follows. Reaction is gauged by public and organizational response to a series of hearings, workshops, correspondence, and private conversations.

The final phase of the rulemaking process takes place after the public comment period has ended. The proposed rule undergoes yet another internal management review, and changes may occur as a result of new information received or the reinterpretation of existing information. Once again, OMB is given an opportunity to comment within a thirty day review period. After the document is signed by the administrator, the final rule is published in the Federal Register. Counting OMB review, an EPA study cited by Bryner found that the average regulation took 33 months to complete.[17] It should be noted that most rules are routinely challenged in court by affected industries or environmental groups, thereby adding months or even years of delay to the beginning date for industry compliance.

EPA decision making on the regulatory front has been complicated by statutory ambiguity, the lack of precedent, technological uncertainty, resource constraints, internal disagreement within the agency over the shape and direction of proposed rules, and congressionally mandated deadlines. The enactment of RCRA by Congress in 1976 resulted in a noncontroversial law that was largely overshadowed by the preoccupation of industry and elected officials with the debate over the Toxic Substances Control Act. As a result, the language of Subtitle C dealing with hazardous waste did not receive the depth of legislative scrutiny characteristic of other pollution-control laws. An unfortunate consequence of the combination of this characteristic and the dearth of state hazardous waste legislation was the creation of a statute delegating substantial discretion to EPA with relatively few substantive guidelines (such as an operational definition of hazardous waste) for the construction of regulations.[18]

A corollary condition that accompanied the lack of statutory precedent was an insufficient body of scientific evidence to develop rules that were sufficiently sound to withstand lawsuits initiated by industry or environmental groups. EPA administrators were confronted at the outset with a serious problem definition: What do we actually mean by labeling something as a hazardous waste? Should a "degree of hazard" approach be developed in lieu of standards classifying wastes as hazardous or nonhazardous? To what extent, if at all, should the concentration of a given substance be taken into account, since some chemicals are not considered hazardous in small amounts or in dilute form? Answers to questions of this sort posed a dilemma for administrators attempting to develop quality standards that were beneficial from an environmental quality or public health perspective without imposing an unduly harsh economic burden on regulated industries.[19]

Another problem with advancing the regulatory agenda early on was attributable to disparate viewpoints espoused by administrators within EPA. Rulemaking responsibilities were initially carved up into separate pieces and assigned to different offices within the agency. This was troublesome, since no one was placed in charge of coordinating the overall shape and direction of the program.[20] The problem was also exacerbated by difficulties with disciplinary communication. Agency lawyers responsible for ensuring that emerging rules were legally defensible found themselves spending a great deal of time rewriting drafts of regulations submitted by engineers and other technical staff.[21] Organizational and disciplinary differences also were reflected in substantive disagreements over issues such as the worthiness of regulatory exemptions for special wastes and small-quantity generators.[22]

Finally, regulatory development was slowed by a combination of resource limitations and congressionally mandated deadlines, which were well-meaning but unrealistic. RCRA and Superfund were enacted within a period of less than four years. Both programs had serious technical problems to begin with, which were compounded by a lack of staff and funding.[23] Fewer than 15 individuals were initially involved in the process of writing RCRA regulations, a task that would ordinarily require four to six times that number of staffers to produce the necessary documents within the allotted time period.[24]

The net effect of the constraints mentioned thus far was to delay the promulgation of rules well beyond the date established by Congress. The last major batch of regulations for RCRA was actually promulgated in May 1980—two years late. Such tardiness was especially frustrating not only to EPA staffers, members of Congress, and environmental leaders but to state officials as well, whose program development efforts were essentially put on hold until they had a set of rules—and accompanying interpretative guidelines—to which to respond.

When Congress reauthorized both RCRA and Superfund in the mid-1980s, EPA was assigned the responsibility of meeting an even greater number of regulatory deadlines. Moreover, a conditional requirement subsequently referred to as the "hammer clause" was inserted into the Hazardous and Solid Waste Amendments of 1984. It specified a rather strict set of standards that would automatically go into effect by a particular date in the absence of agency progress in crafting these rules.

A change of this magnitude clearly influenced one aspect of the relationship between regulated groups and EPA. From the perspective of the Chemical Manufacturers' Association and other trade organizations, the incentive structure changed rather abruptly from a delay-inducing strategy to one of encouraging a speed-up of rulemaking efforts. The prospect of coping with EPA standards developed expeditiously with industry input has been viewed in a more favorable light by these parties than the stricter rules scheduled to come on-line by a particular date. Despite such prods, progress has been limited. A GAO report revealed that by April 1988, 66 of the 76 deadlines had come due and that EPA had completed action on fewer than half of these. Priority was assigned to program elements containing hammer provisions (three of four completed).[25]

Permits

While rulemaking has occupied a significant portion of the time and attention devoted to RCRA by EPA officials, it is not the only important form of regulatory activity undertaken by agency administrators. They also share responsibility with state officials for the issuance of permits required of tsd facilities. Ideally, this should be a priority concern, since administrators have an opportunity to shape the behavior of regulated entities early on by specifying the conditions under which they may operate. In practice, this activity has been complicated by the sheer number of parties that are subject to RCRA requirements, the technical complexity of the permitting process, and the realization that an overly strict set of conditions may produce the unintended effect of putting facilities out of business and/or increasing the temptation for some firms to engage in illegal waste-disposal practices.

The RCRA permit was designed to ensure that any tsd facilities constructed after November 19, 1980 (the effective date of regulations specifying technical standards for hazardous waste disposal and facility construction), meet new environmental quality standards. Since EPA had neither the time nor the resources to obtain and verify the information needed to issue permits for all relevant facilities, a special category of "interim status" permits was created. This allowed uninspected landfills to continue receiving hazardous wastes with the proviso that facility operators would submit an application for the RCRA permit and would strive to meet the newly established environmental quality standards.[26]

A monkey wrench of sorts was tossed into a permitting process already complicated by an administrative backlog of facility applicants when Congress enacted HSWA in 1984. Interim-status land-based facilities were given a year to demonstrate compliance with groundwater monitoring and financial responsibility requirements. Failure to provide documentation of compliance would result in a loss of eligibility to continue handling wastes.

The effects of this change were twofold. First, much of the nation's capacity to legally store or dispose of hazardous waste in land-based facilities was lost. Nearly two-thirds of tsd facilities with interim-status permits were unable to comply.[27] Second, despite stepped-up efforts by EPA officials to conduct inspections and to enforce regulatory requirements, actual compliance rates at 1,450 land disposal facilities between 1985 and 1987 were not very impressive—around 50 percent.[28]

ENFORCEMENT

While EPA's responsibility for the implementation of both RCRA and Superfund is well established, its role in the day-to-day enforcement activities has varied. Federal officials are more likely to take part in the administrative details of Superfund, while state officials are the primary enforcers of RCRA. Both are fairly complex as regulatory programs go, and the process of achieving the desired programmatic outcome will often involve the participation of myriad policy actors from industry and all levels of government.

How Enforcement Works

How is Superfund implemented? A summary of the enforcement process is presented in Figure 4–1. EPA ranks hazardous waste sites on the basis of scores received on the Hazardous Ranking System (HRS), an index designed to measure the severity of risks to humans or the environment posed by exposure to the contaminated area. Sites ranking high on this index are considered prime candidates for inclusion on the NPL.[29] There are currently 1,189 sites on the list, a figure that includes both federal and nonfederal properties. Smaller sites probably will be handled at the state level (most have Superfund programs to handle abandoned dump sites that are ineligible for federal funding).

Enforcement actions are coordinated through EPA's Office of Solid Waste and Emergency Response (OSWER), Office of Enforcement and Compliance Monitoring (OECM), and ten regional offices. While OECM is responsible for providing for direction and guidance on Superfund enforcement actions to OSWER and the regions, it is the regional administrators who actually prepare and carry out cleanup actions at NPL sites. Their tasks are facilitated by a variety of specialists, including site coordinators who oversee removal actions, attorneys responsible for the legal details to be incorporated within settlements and enforcement actions, civil investigators who seek out potentially responsible parties (PRPs) in their assessment of financial liability, and project managers whose main focus is the technical aspects of enforcement cases.[30]

EPA is responsible for two types of site cleanup—emergency actions and remedial actions. The former are short-term responses undertaken to address immediate threats to public health. Cleanups of this sort are not intended to be long-lasting. Remedial actions are long-term efforts to mitigate or eliminate hazardous conditions at an NPL site. The process begins with a remedial investigation and a feasibility study (RI/FS) conducted by EPA to determine the type and volume of waste at the site and the most appropriate remedy among several considered options to achieve cleanup goals.

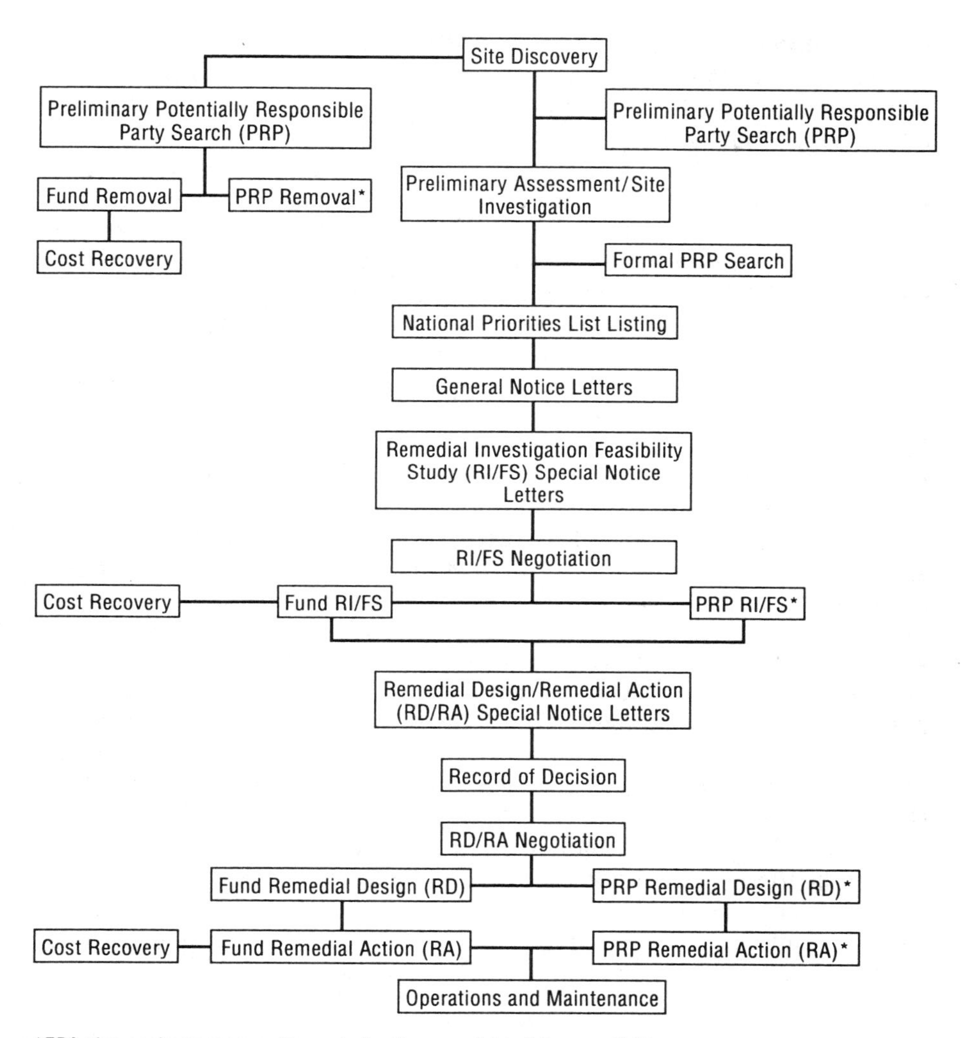

*EPA also seeks to recover its costs for the oversight of these activities.

FIGURE 4–1 Superfund Cleanup and Enforcement Process

Source: General Accounting Office, *Superfund: A More Vigorous and Better Managed Enforcement Program Is Needed.* GAO/RCED-90-22, December 1989, p. 13.

A parallel action occurring simultaneously with the RI/FS study is the search for PRPs. Only then can EPA officials establish which firms, individuals, or other entities are responsible for polluting a particular site and whether any or all of these parties are financially able to contribute to the cost of ameliorating the problem. PRPs are contacted and informed about their potential liability. Those who disagree with the allegation that they have disposed of pollutants at a given site may choose to contest such charges in court. However, this may prove to be an overly risky strategy for some, since EPA is empowered to collect treble damages from unsuccessful PRP plaintiffs.[31]

If PRPs agree to participate in site cleanup costs, EPA will negotiate an agreement that is built into a consent decree. It can include settlements calling for most or all of these costs to be borne by responsible parties; a mixed-funding approach, which finances cleanup actions from monies collected from the Superfund and responsible parties; or a *de minimis* agreement, which eliminates from the negotiating process parties that have contributed small amounts of less toxic waste to a site. The latter agreement represents a potentially useful way to save money and staff resources in site negotiations involving a large number of PRPs.

PRPs have ample reason to bargain in good faith because the federal government has been given substantial enforcement authority under Superfund. If negotiations break down, EPA officials have several options. Administrators can undertake relatively low-level actions, such as obtaining reimbursement costs for cleanup actions carried out by the agency or a court order requiring PRPs to do the work themselves, or a more aggressive response, such as the issuance of an administrative order. The latter approach compels the responsible party to perform the work immediately or face the imposition of sanctions. Failure to comply with the terms of an administrative order can result in fines of $25,000 per day and punitive damages amounting to three times the cost of site cleanup if EPA does the job.[32]

Once EPA has determined that a site no longer poses a threat to human health or environmental quality, the site can be *de-listed*; that is, removed from the NPL. This decision is reached after consulting with state officials and holding a public comment period. To date, relatively few sites (33 as of February 1991) have been de-listed.

The process of enforcing RCRA is equally complex (see Figure 4–2). It should be noted that EPA plays a smaller role in the intergovernmental assignment of policy responsibilities than is evident with Superfund. Most states have primacy and are therefore responsible for the day-to-day management of this program, which involves the "cradle-to-grave" regulation of hazardous wastes. EPA's primary tasks include inspecting waste handlers, issuing permits for tsd facilities, monitoring state enforcement actions, and, if necessary, instigating enforcement proceedings against those in violation of the law. Tasks such as inspections, issuance of permits, and enforcement are shared with state officials.

Periodic inspection of hazardous waste facilities is the first step. EPA guidelines developed in 1980 (incorporated within HSWA in 1984)

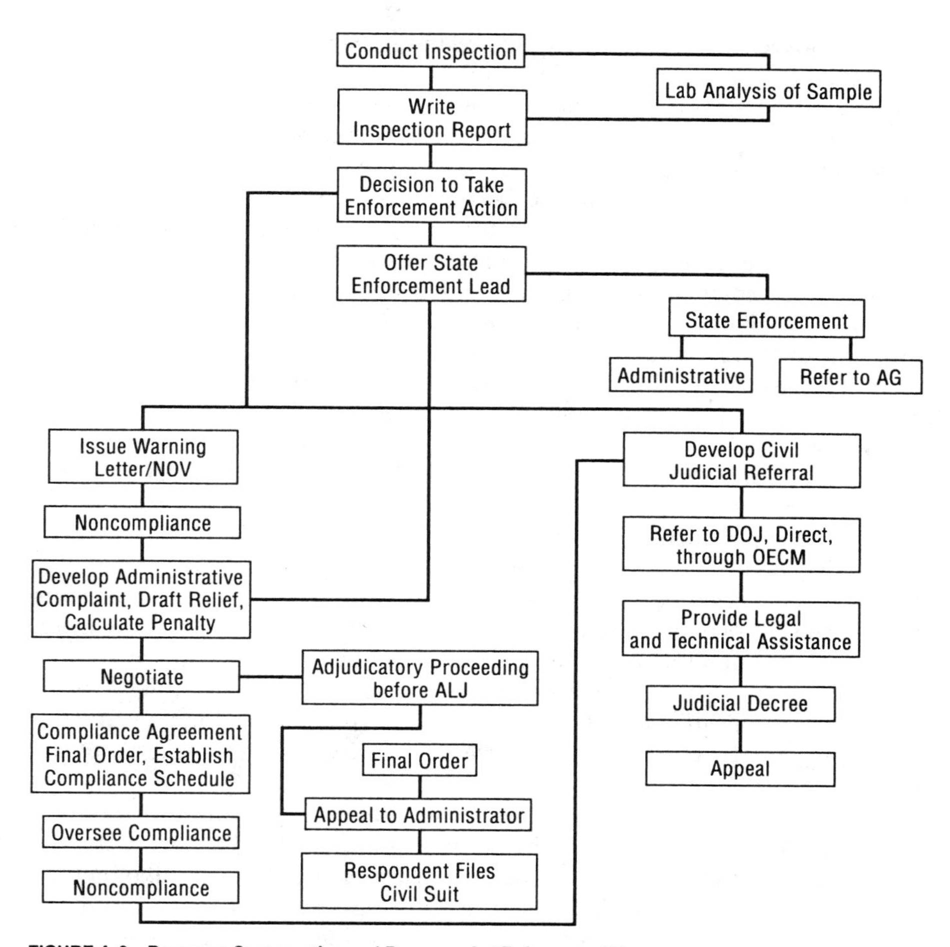

FIGURE 4–2 Resource Conservation and Recovery Act Enforcement Process

Source: Environmental Protection Agency, *The Nation's Hazardous Waste Program at a Crossroads: The RCRA Implementation Study* (Washington, D.C.: Government Printing Office, July 1990), p. 59.

called for each interim-status or permitted tsd facility to be visited at least once every two years. Inspections are supposed to be carried out more frequently for facilities that store wastes or those that are operated by state or federal agencies. Site visitation also can be triggered by citizen complaints. Violations are noted by inspectors and communicated to facility managers with the expectation that corrective actions will be taken in an expeditious manner. Regulatory officials usually try to resolve problems through using letters, meetings, or phone calls before proceeding to a more formal phase of enforcement.[33]

If individuals or organizations fail to respond to the problems cited in the inspection report, agency administrators can consider several options. For relatively minor offenses, a preferred strategy is to send the offending party a warning letter or notification of violation (NOV), which identifies the problems in need of attention and specifies a target date for corrective actions to be taken. Offenders' failure to respond to NOVs in a timely fashion (within 60 days) will lead regulatory officials to up the ante by imposing administrative orders or referring cases to the state attorney general for prosecution.[34]

Most serious problems, such as violations posing a direct and immediate threat to public health or environmental quality, are handled by issuing an administrative order. The nature of the offense and compliance deadlines are spelled out here (as with NOVs), but the order is backed up with sanctions in the form of fines or imprisonment. State laws vary, but federal policy permits fines as high as $25,000 per day for particularly egregious offenses to be levied against polluting entities by EPA officials until the problems have been corrected.

EPA has developed guidelines for calculating the magnitude of financial penalties based on a number of factors. Chief among these is the degree of economic benefit received by the polluting organization from its refusal to comply with RCRA standards. Other factors taken into consideration are the financial health of the firm or agency, prior citations for pollution-control violations, a good-faith effort to ameliorate problems by the facility owner/operator, and other extenuating circumstances.[35]

Criminal prosecution of those who generate or handle hazardous waste in an unsafe manner may be pursued if it can be demonstrated that EPA regulations were intentionally disregarded. Examples include illegal disposal practices (such as the incineration of wastes without the necessary permits); failure to comply with reporting requirements for the production, transport, storage, or disposal of hazardous wastes; destruction or alteration the contents of industry manifests; and the export of wastes without the consent of the appropriate governmental representatives of the receiving country. The U.S. Department of Justice can seek jail terms of up to five years for serious offenders, and even stiffer sentences can be imposed for repeat violations. It is important to note that polluting firms can be sued not only by relevant public authorities for noncompliance with RCRA but by any citizen or environmental organization as well (absent any ongoing enforcement actions).[36]

One other aspect of the program in place since the enactment of HSWA in 1984 is the requirement that corrective action be taken by permitted facilities to clean up existing or past releases of hazardous wastes. The importance of this requirement is reflected in EPA estimates indicating that the number of facilities in need of corrective action is three times greater than the number of NPL sites.[37] Facility operators are responsible for investigating the magnitude of the problem as well as designing and implementing a management solution. Either EPA or state officials can be in charge of project oversight, although in practice federal officials have undertaken the lion's share of supervisory activities.[38]

Enforcement Results

Let us now turn from a discussion of enforcement processes to an analysis of results. The implementation of Superfund between 1981 and 1983 dovetails with the foci on deregulation and economic growth during the first term of the Reagan Administration and was closely intertwined with the tumultuous reign of EPA Administrator Burford and Assistant Administrator Lavelle. Allegations of political manipulation of the Superfund program involving such questionable practices as the negotiation of "sweetheart deals" on cleanup liability between agency operatives and industry officials in advance of data showing actual cleanup costs prompted a spate of congressional oversight hearings (discussed earlier) and cast a shadow on EPA efforts to promote a less adversarial relationship between the agency and regulated industries in program management decisions.

As a result of political problems coupled with the start-up problems typically associated with the development of new, complex programs, EPA made little progress early on in cleaning up NPL sites. By mid-l983, only two sites had been cleaned up, while the planning process had been initiated for long term cleanups on 23 other sites.[39] Greater emphasis was placed on emergency response actions, and EPA's performance under this criterion was somewhat better. Eighteen initial remedial actions, two planned removals, and 107 immediate removals were completed within the same time period.[40] According to then–Superfund Director William Hedeman, program spending was deliberately slowed to demonstrate a lack of need for its subsequent reauthorization.[41]

When Ruckelshaus took over as EPA administrator in mid-1983, the agency enforcement strategy changed. Burford had focused on private-party cleanup as the point of departure, an approach that emphasized budgetary savings over the pace of cleanup actions. Ruckelshaus adopted a more aggressive enforcement posture, exemplified in part by his willingness to use lawsuits to produce increased industry compliance with Superfund. A 1985 GAO study revealed that Superfund-financed remedial activity was occurring at 298 of the 538 sites on the NPL, although the vast majority of actions could be described as part of the feasibility study phase rather than actual cleanup actions.[42] Indeed, as Bowman notes, the dearth of spadework can be attributed to the expense and time-consuming nature of preliminary decisions of this sort. The average RI/FS study takes a year-and-a-half to complete and costs $800,000.[43]

Ruckelshaus's philosophical stance toward Superfund enforcement was continued by his successor, Lee Thomas, between 1985 and 1989 and, more recently, by the current EPA administrator, William Reilly. An accelerated pace is reflected in several areas of activity between 1983 and 1987, including the number of RI/FS studies launched (from 169 to 494), the number of sites at the remedial design stage (from 17 to 94), the initiation of remedial actions (from none to 89), and the number of permanent cleanups (29) completed.[44] Contributing to a more favorable environment for enforcement was the enactment in 1986 of SARA (see Chapter 2), which incorporated a goals and timetables approach for cleanup actions.

While a modicum of progress has been made over the past several years, a number of critical studies have examined EPA's management of the Superfund program.[45] Some of these studies address procedural concerns such as the degree of effort devoted to the search for PRPs (to be discussed in the concluding chapter). Others focus on institutional or resource factors responsible for impeding or accelerating policy outputs, notably emergency response actions or permanent cleanups.

Many of the problems associated with Superfund enforcement in the early 1980s were attributed to the policy priorities of the Reagan Administration and the belief among top administrators that environmental protection goals should be subordinated to economic expansion and deregulation.[46] Thus, early on, a number of factors—such as a lack of program commitment among EPA officials, a lack of financial and staff resources, and the intergovernmental communication of policy objectives—contributed to a slow start in cleanup actions as well as the inevitable delay caused by inexperience.

Other hurdles have been identified by Bowman in several analyses of Superfund implementation.[47] They include the complexity of interagency and intergovernmental decision-making responsibilities, unintended consequences of decision rules adopted by EPA, and the unwillingness of many PRPs to accept financial responsibility for all or part of site cleanup costs. The latter problem can be illustrated by EPA's experience in dealing with 12 PRPs at the Bluff Road NPL site in South Carolina. While seven of the waste generators chose to cooperate and eventually settled on a payment of $1.95 million, the remaining amount of site cleanup costs, or 25 percent of the total, was financed by monies drawn from the Superfund account. The nonparticipating entities were eventually sued by the federal government.[48]

More recent accounts of Superfund implementation indicate that some of the earlier problems, such as funding, procedural consistency in site project decision making, political intervention, and administrative commitment, have been resolved or at least addressed.[49] Reilly sought to make Superfund enforcement a priority concern after his appointment as EPA administrator in 1989. A careful study by Hird suggests that project site decision making by EPA officials is more likely to reflect sensitivity to the technical requirements of site cleanup (as measured by Hazardous Ranking System scores) than political influence exercised by members of Congress belonging to relevant oversight committees or appropriations subcommittees.[50] Nevertheless, progress measured in terms of permanent cleanups is unlikely to be achieved as quickly as many elected officials would prefer because of the procedural complexity of NPL site decisionmaking and the failure of EPA to provide clear and consistent guidelines for the establishment of priorities among sites and the determination of how clean is clean enough.[51]

Turning now to RCRA implementation, it should be noted that EPA plays a smaller role in the intergovernmental assignment of policy responsibilities here than is evident with Superfund. Most states have achieved primacy for the basic (i.e., pre-HSWA) regulatory program and are therefore responsible for day-to-day management activities associated with the "cradle-to-grave" administrative scheme. EPA's primary tasks include the inspection of waste handlers, the issuance of permits for tsd facilities, the monitoring of state enforcement actions, and, if necessary, the instigation of enforcement proceedings against those in violation of the law. Some of these tasks, notably inspections and the issuance of permits and enforcement, are shared with state officials (see Chapter 5).

Early experience with the implementation of RCRA was disappointing to many who saw this program as the cornerstone of policy efforts to control hazardous wastes. EPA officials were confronted with many of the same problems encountered with Superfund—notably political problems, technological complexity, and resource limitations. Putting the regulatory apparatus in place so that state officials could begin enforcing the program proved to be especially frustrating.

EPA's rulemaking strategy under RCRA was to seek agreement among interested parties through public participation and negotiations prior to the promulgation of final rules. The interests of EPA, waste-generating firms, waste-disposal companies, state policymakers, and environmentalists were sufficiently diverse to greatly complicate the attainment of a policy consensus.[52] The rulemaking process spanned seven years (1976–1983) and eventually resulted in the production of six extensive regulations and over 40 amendments and clarifications.[53]

Because the rules were not yet in place, it is hardly surprising that EPA encountered implementation problems similar to those found with the Superfund program. The lull in RCRA enforcement and its subsequent rise after Congress issued a contempt citation against EPA Administrator Burford in the midst of the political imbroglio over Superfund are documented in a paper by Wood and Tzoumas. In several charts that track the monthly flow of hazardous waste administrative actions between January 1981 and September 1988, the authors demonstrate that EPA did not undertake many inspections of waste handlers or issue warning letters to noncompliant parties until October 1983—nearly a year after the controversy was resolved.[54]

Despite EPA's inability to implement policy through "normal" enforcement procedures, Wood and Tzoumas suggest that agency professionals did not throw in the towel out of frustration with political and budgetary constraints but instead opted to follow a different administrative path. The most cost-effective and visible means of enforcement available to EPA administrators was the stepped-up use of litigation. In this way, compliance with hazardous waste laws could be increased by presenting a more credible regulatory posture. This was achieved by increasing the number of administrative orders and financial penalties in relation to inspection.[55]

Beginning in 1983, greater emphasis has been placed on enforcement actions within EPA and authorized state agencies. Both staff and fiscal resources have increased and Congress strengthened EPA's authority to take action against polluting firms under HSWA. Meanwhile, programmatic changes also were made to improve agency effectiveness. An Enforcement Response Policy calling for greater differentiation between classes of regulatory violators was developed, thereby aiding administrators in the task of setting enforcement priorities. It targeted high-priority violators for formal enforcement action involving administrative orders or penalties while less egregious offenders were placed in an enforcement category that made greater use of warning letters or notifications of violation (NOV).[56]

Has the new policy made a difference in the EPA's ability to enforce the law in a timely and appropriate fashion? This question was addressed in a GAO report that analyzed federal enforcement data collected between October 1985 and July 1987 in Regions II, V, and VI. The report was critical of EPA's performance, noting that an appropriate enforcement action was selected in approximately two-thirds of all cases reviewed, while only about one-third of enforcement decisions were made within the suggested time guidelines.

In addition, these figures mask some problems in the distribution of actions taken. EPA was much less likely to handle high-priority violators appropriately and expeditiously than lower-priority violators, since it is much easier to issue a warning letter than to prepare the necessary documentation for an administrative order with a penalty. These problems have been attributed to both resource constraints and disagreement among administrators over the appropriateness of using a particular enforcement approach under varied conditions.[57]

EPA efforts to implement the corrective action requirements spelled out in HSWA have not yet produced much in the way of results, although the process is underway. Agency officials estimate that approximately 1,600 facilities nationwide have completed the assessment phase, which identifies actual or potential releases of hazardous waste at every waste management unit. Their data also indicate that most of these (about 80 percent) subsequently require an RFI (RCRA Facility Investigation).[58] This is a more systematic analysis of the amount of waste released and the rate at which it is progressing through the ground.

Thus far, relatively few RFIs have been approved or completed, although some cleanup actions have been taken at facilities that have not finished preliminary studies. In part, the delay can be attributed to the nature of the process. The gathering and processing of information about waste releases is inherently time-consuming. Second, few states have shown any interest in obtaining primacy for corrective action responsibilities because of the fear of escalating cleanup costs. Third, EPA has yet to receive resource commitments that are commensurate with the task at hand. Within EPA headquarters, 17 work years have been budgeted for RCRA corrective action activities, compared with 275 work years for Superfund.[59] In short, much remains to be done.

Another programmatic responsibility assigned to EPA by Congress under RCRA is the oversight of state enforcement actions. If state officials are unable or unwilling to implement the law, EPA officials are obliged to initiate enforcement proceedings, or, in extreme cases, to preempt state regulatory authority. The GAO has again faulted EPA for a less than vigilant regulatory posture. For example, action was not taken in any of the cases in which states failed to meet the decisional criteria established in the Enforcement Response Policy. A survey of EPA officials revealed that their reluctance to intervene could be attributed to a fear of jeopardizing working relationships between regional and state administrators, resource limitations, and the belief that, in some cases, the states were making reasonable progress in bringing parties into compliance.[60]

CONCLUSIONS

There are several types of decisions made by EPA that shape the implementation of hazardous waste programs. The ability and willingness of the administrator to communicate policy objectives to regional staff, state program administrators, and industry officials clearly sets the tone for enforcement decisions down the line. Two EPA administrators who have given higher priority to a media-based strategy designed to increase voluntary compliance with RCRA and Superfund are William Ruckelshaus and William Reilly.

EPA rules and procedures also affect the administration of hazardous waste programs. The delegation of program management authority to the states, the promulgation of regulations, and the issuance of permits require decisions by administrators that carry important consequences for the attainment of policy goals. This is particularly true for rulemaking. The uncertainty created by overly broad statutory guidelines is reduced by specifying not only the substantive interpretation of policy but also the parties likely to be affected, the officials in charge, the range of permissible actions that can be taken, criteria used and the relative weight assigned to each, and the deadlines for the completion of a given task.

The application of rules and procedures to specific cases by EPA officials has been complicated and in some cases derailed by several factors. The delegation of program management authority to the states was virtually complete by the mid-1980s for the basic RCRA regulatory package but has been proceeding at a much slower pace since the enactment of HSWA. Few states have achieved primacy for the newer policy because of the requirement that state regulations be substantially equivalent to federal regulations, a decision rule that effectively reduced the amount of discretion available to state administrators to try new initiatives.

Early EPA action on rulemaking following the adoption of RCRA in 1976 was stifled by several factors, including a lack of precedent, statutory ambiguity, internal disagreement within the agency on how to proceed, and resource constraints. Since the mid-1980s, things have improved in terms of regulatory development, in part because of improved ability to operate within a diverse organizational environment, somewhat better funding, and the need to meet an increasingly ambitious set of goals within deadlines established by Congress. The issuance of permits for tsd facilities has followed a parallel course, since the development of standards precedes waste management activities. In the past few years, considerable progress has been made in eliminating the backlog of permit applications.

The most direct form of implementation carried out by EPA is enforcement. Following a period in the early 1980s of benign neglect and underfunding for enforcement activities, EPA administrators William Ruckelshaus and Lee Thomas attempted to place greater emphasis on deterrence as a regulatory strategy. They succeeded in obtaining an increased budget for enforcement actions and gave notice that noncompliant firms would be expected to follow the rules or be fined or prosecuted. This philosophy also has been carried forward by President Bush's EPA administrator, William Reilly, and the results are reflected in a steady increase in both enforcement actions and lawsuits against polluting firms.

In short, EPA responsibilities for communicating policy objectives, writing regulations, issuing permits, and enforcing regulations, have evolved considerably since the early 1980s as a consequence of changing policy and administrative priorities in the administrator's office. Both programs, particularly RCRA, required adaptive learning and behavior not only from EPA but from administrators working for state-approved environmental agencies as well. Both resource levels and procedural factors affect the likelihood of implementation, but the most important impetus for improving program performance is the appointment of an administrator with a commitment to hazardous waste policy goals and good communication skills.

Overall, EPA has managed to make slow but steady progress in increasing the number of enforcement actions and NPL site cleanups. However, this is not to say that the desired state of affairs has been achieved from an agency perspective. A number of studies have identified continuing trouble spots in the administration of both RCRA and Superfund and have offered a host of policy and managerial changes to improve organizational performance.

NOTES

1. Alfred Marcus, *Promise and Performance: Choosing and Implementing an Environmental Policy* (Westport, Conn.: Greenwood Press, 1980), pp. 88–90.

2. This is consistent with other actions taken by Ruckelshaus to accentuate the importance of regulatory credibility, such as his belief that EPA could act as a type of regulatory backstop for state environmental agencies.

3. Among others, consult Malcolm Goggin, Ann O' M. Bowman, James P. Lester, and Laurence J. O'Toole, Jr., *Implementation Theory and Practice* (Glenview, Ill.: Scott, Foresman/Little, Brown, 1990), especially Chapter 3; and Joseph Di Mento, "Can Social Science Explain Organizational Noncompliance with Environmental Law?" *Journal of Social Issues*, *45*, No. 1 (1989), pp. 109–30.

4. Ibid.

5. Rochelle Stanfield, "Ruckelshaus Casts EPA as 'Gorilla' in States' Enforcement Closet," *National Journal* (May 26, 1984).

6. Richard Waterman, "Reagan and the EPA: Revolution and Counterrevolution," Midwest Political Science Association Conference Paper, Chicago, Illinois, 1990, p. 16.

7. Michael D. Reagan, *Regulation: The Politics of Policy* (Boston: Little, Brown, 1987), Chapter 8; Harvey Lieber, "Federalism and Hazardous Waste Policy." In James P. Lester and Ann O' M. Bowman, eds., *The Politics of Hazardous Waste Management* (Durham, N.C.: Duke University Press, 1983), pp. 60–72; and Ann O'M. Bowman, "Hazardous Waste Management: An Emerging Policy Area within an Emerging Federalism," *Publius*, *15* (Winter 1985), pp. 131–44.

8. Cheryl H. Wilf, "Administrative Preemption & State Authority: Revisiting the Middle Ground Thesis," American Political Science Association Conference paper, San Francisco, 1990.

9. Stanfield, p. 1035.

10. Anne M. Burford, *Are You Tough Enough?* (New York: McGraw-Hill, 1986), p. 56.

11. Michael Fix, "Transferring Regulatory Authority to the States." In George C. Eads and Michael Fix, eds., *The Reagan Regulatory Strategy: An Assessment* (Washington, D.C.: Urban Institute, 1984), pp. 153–79.

12. Stanfield, p. 1035.

13. Keystone Center, *State–Federal Relations in the RCRA Regulatory Program* (Keystone, Colo.: Keystone Center, June 1989).

14. Ibid.

15. Michael Kraft, Bruce Clary, and Richard Tobin, "The Impact of New Federalism on State Environmental Policy." In Peter Eiseinger and William Gormley, eds., *The Midwest Response to the New Federalism* (Madison: University of Wisconsin Press, 1988), pp. 204–33.

16. Ann Bowman, "Superfund Implementation: Five Years and How Many Cleanups?" In Charles Davis and James P. Lester, eds., *Dimensions of Hazardous Waste Politics and Policy* (Westport, Conn.: Greenwood Press, 1988), p. 139.

17. Gary Bryner, *Bureaucratic Discretion: Law & Policy in Federal Regulatory Agencies* (New York: Pergamon Press, 1987), p. 99.

18. Marc K. Landy, Marc J. Roberts, and Stephen R. Thomas, *The Environmental Protection Agency: Asking the Wrong Questions* (New York: Oxford University Press, 1990), pp. 94–96.

19. Ibid. See also Sam Carnes, "Confronting Complexity and Uncertainty: Implementation of Hazardous Waste Management Policy." In Dean Mann, ed., *Environmental Policy Implementation* (Lexington, Mass.: Lexington Books, 1982), pp. 35–50.

20. Ibid., pp. 101–6.

21. Ibid., p. 103.

22. Ibid., p. 106.

23. Walter Rosenbaum, *Environmental Politics and Policy*, 2nd ed. (Washington, D.C.: CQ Press, 1991), p. 204.

24. Landy, et al., *The Environmental Protection Agency*, p. 103.

25. General Accounting Office, *Hazardous Waste: New Approach Needed to Manage the Resource Conservation and Recovery Act* (GAO/RCED-88-115, July 1988), p. 3.

26. Caroline Wehlin, "RCRA Permitting," *Natural Resources and Environment*, 2, No. 3 (Winter 1987), pp. 27–30, 53–55.

27. Ibid., p. 29.

28. GAO, *New Approach Needed to Manage RCRA*, p. 4.

29. General Accounting Office, *Superfund: A More Vigorous and Better Managed Enforcement Program Is Needed* (GAO/RCED-90-22, December 1989), p. 11.

30. Ibid., p. 17.

31. Ibid., pp. 14-15.

32. Ibid., p. 16.

33. Richard Fortuna and David Lennett, *Hazardous Waste Regulation: A New Era* (New York: McGraw-Hill, 1987), especially Chapter 13.

34. Ibid.

35. Ibid.

36. GAO, *New Approach Needed to Manage RCRA*.

37. U.S. Environmental Protection Agency, *The Nation's Hazardous Waste Program at a Crossroads: The RCRA Implementation Study* (Washington, D.C.: Government Printing Office, July 1990), Chapter 7.

38. Ibid.

39. Steven Cohen, "Defusing the Toxic Time Bomb: Federal Hazardous Waste Programs," in Norman Vig and Michael Kraft, eds., *Environmental Policy in the 1980s* (Washington, D.C.: CQ Press, 1984), p. 287.

40. Ibid.

41. Ibid., p. 288.

42. Cited in Bowman, "Superfund Implementation," p. 136.

43. Ibid., p. 138.

44. James P. Lester, "Implementing Intergovernmental Regulatory Policy: The Case of Hazardous Waste." In Shyarnal Majumdar, E. Willard Miller, and Robert Schmalz, eds., *Management of Hazardous Wastes* (Phillipsburg, N.J.: Pennsylvania Academy of Science, 1989), pp. 313–28.

45. Examples include Office of Technology Assessment, *Are We Cleaning Up? 10 Superfund Case Studies* (Washington, D.C.: Government Printing Office, June 1988); General Accounting Office, *Superfund: Interim Assessment of EPA's Enforcement Program* (GAO/RCED-89-40BR, October 1988); John Hird, "Superfund Expenditures and Cleanup Priorities," *Journal of Policy Analysis and Management, 9* (Fall 1990), pp. 455–83; and James P. Lester, "Superfund Implementation: Exploring the Conditions of Environmental Gridlock," *Environmental Impact Assessment Review, 8* (1988), pp. 63–70.

46. Michael Kraft and Norman Vig, "Environmental Policy from the Seventies to the Nineties." In Norman Vig and Michael Kraft, eds., *Environmental Policy in the 1990s* (Washington, D.C.: CQ Press, 1990), p. 15.

47. Bowman, "Superfund Implementation," pp. 129–46; "Explaining State Response to the Hazardous Waste Problem," *Hazardous Waste, 1*, No. 3 (1984), pp. 301–8; and "Intergovernmental and Intersectoral Tensions in Environmental Policy Implementation: The Case of Hazardous Waste," *Policy Studies Review* (November 1984), pp. 230–44.

48. Bowman, "Superfund Implementation," p. 142.

49. GAO, "Superfund: A More Vigorous Enforcement Program Is Needed;" and Center for Hazardous Waste Management, *Coalition on Superfund Research Report* (Chicago: Illinois Institute of Technology, September 1989).

50. Hird, "Superfund Expenditures and Cleanup Priorities," pp. 477–78.

51. OTA, *Are We Cleaning Up?*, pp. 1–4; and James Lester, "Implementing Intergovernmental Regulatory Policy."

52. Malcolm Getz and Benjamin Walter, "Environmental Policy and Competitive Structure: Implications of the Hazardous Waste Management Program," *Policy Studies Journal 9* (Winter 1980), pp. 404–14.

53. Charles Davis, "Implementing the Resource Conservation and Recovery Act of 1976," *Public Administration Quarterly* (Summer 1985), pp. 218–36.

54. B. Dan Wood and Kelly Tzoumas, "Politics, Bureaucracy and Hazardous Waste Regulation," Southwestern Social Science Association Conference paper, Little Rock, Arkansas, 1989, p. 10.

55. Ibid.

56. General Accounting Office, *Hazardous Waste: Many Enforcement Actions Do Not Meet EPA Standards* (GAO/RCED-88-140, June 1988).

57. Ibid.

58. EPA, *The Nation's Hazardous Waste Program at a Crossroads*, p. 76.

59. Ibid., p. 78.

60. GAO, *Hazardous Waste: Many Enforcement Actions Do Not Meet EPA Standards*, Chapter 2.

5

HAZARDOUS WASTE POLICYMAKING WITHIN STATE AND LOCAL GOVERNMENTS

State and local governments participate in the policymaking process, but in ways that are circumscribed somewhat by prior federal initiatives. Within the confines of RCRA and Superfund policy guidelines, state officials have latitude to tailor programs that can be matched to jurisdiction-specific problems. Important programmatic responsibilities such as the siting of tsd facilities, the cleanup of abandoned dump sites to qualify for NPL designation, and RCRA enforcement clearly fall within the policy domain of state government, while local governments also may choose to regulate toxic wastes within their borders, independently or in response to state mandates.

With acknowledgment of the importance of EPA, Congress, and other political actors in shaping the general direction of hazardous waste policymaking, there is nevertheless considerable variation in the content of state programs and the stringency of regulatory efforts. I begin with a discussion of policy and regulatory trends. States differ in terms of both program stringency and form of control (e.g., the use of economic incentives as a means of supplementing direct regulatory controls), policies affecting the mitigation or cleanup of abandoned dump sites, facility siting practices, and the involvement of local government in pollution-control efforts. Next, the allocation of own-source monies to address these concerns is examined. A combination of policy and resource commitments provides us with a rough approximation of state institutional capacity for the management of hazardous waste problems.

Third, actions taken by state administrators to implement toxic waste programs, particularly RCRA, are analyzed. My concern in this section goes beyond an aggregate description of state enforcement efforts to include an explanation of between-state differences. State-level indicators such as the severity of pollution problems, the degree of legislative support for prior environmental policy issues, the political strength of industry or environmental groups, and economic factors such as per capita income or dependence on pollution-generating firms as a source of employment are incorporated into the analysis. From a public-policy perspective, research that demonstrates that state enforcement decisions can be attributed to economic, technical, or political factors tells us something about the current intergovernmental assignment of responsibilities. On balance, have state and local governments acted as facilitators or as barriers to the effective implementation of hazardous waste programs?

POLICY TRENDS

The adoption of hazardous waste policies by state and local governments can be attributed, in part, to a top-down influence; namely, the reaction of public officials to the enactment of RCRA and Superfund. However, it is also likely that between-state differences will emerge because of other pressures or factors that are relevant for some states but not for others. We might ask, for example, whether the strength of state hazardous waste laws can be explained by the demographic characteristics of a given jurisdiction, political factors, or situational concerns such as the magnitude of waste disposal problems.

One of the earlier studies focused on the relative impact of technological pressures (or the severity of within-state pollution problems), economic resources, political demands, and organizational autonomy on the development of state hazardous waste legislation. The best predictors were found to be technological pressures, legislative professionalism, and the presence of a single agency to coordinate hazardous waste programs. Thus, states generating more waste tend to enact more stringent laws, as do states with a consolidated environmental bureaucracy.[1]

The authors also discovered that varied circumstances lead to shifts in the explanatory importance of particular independent variables. One is the severity of hazardous waste problems within a state. Elected officials representing jurisdictions plagued with more serious pollution problems tend to assign a larger share of policymaking responsibilities to the legislature than to the bureaucracy. In addition, differing policy outcomes are more pronounced in some regions than in others. If southern states are excluded from analyses of state-level behavior, economic factors such as per capita income can better account for variation in the stringency of state hazardous waste policies. Thus, the severity of waste management problems and organizational characteristics facilitate hazardous waste policymaking across all states, while political and economic factors are useful predictors under more limited circumstances.[2]

Other writers suggest that the adoption of strict hazardous waste laws is a function less of problem severity or extent of organizational arrangements than of the economic importance of pollution-generating firms to the state. Here, the relevant question for state policymakers is how public health or environmental quality can be reconciled with economic development objectives. A number of studies indicate that the latter objectives often take precedence over the former. Since states compete for these industries, policymakers are reluctant to impose strict regulatory standards that would place companies operating within their borders at a competitive economic disadvantage with similar companies in neighboring states. A consequence is the enactment of state laws designed to meet minimum federal standards in ways that are less costly to industry.[3]

How does this occur? One assumption that has subsequently been validated empirically is that states with a higher percentage of economic activity related to the manufacture of chemicals or other waste-producing products are less likely to favor the adoption of tough regulatory laws. Yet some of these states have been quite sensitive to a wide array of environmental concerns over the years (New Jersey and California exemplify the model of states that have successfully retained a commitment to strong environmental laws without jeopardizing their economic base).

Admittedly, policymakers operating in highly populous states recognize that industry officials will accept higher regulatory costs to some degree because of easier access to markets and skilled labor.[4] But another factor to consider is the socialization of cost thesis. This refers to the argument that regulatory compliance costs are in effect shouldered by government rather than the private sector.[5] Elected officials in affluent states are especially supportive of hazardous waste policy initiatives, since the benefits of attaining pollution-control objectives can be achieved without suffering the economic or political cost of industry relocation. Fiscal options such as governmental subsidies for innovative pollution-control strategies or equipment are clearly more politically viable in affluent states than in poorer jurisdictions.[6]

Other factors contributing to state hazardous waste policymaking have been revealed in case analyses of specific states. Interest groups have achieved some success in pushing for pollution-control legislation, especially in states where outdoor recreation or scenic values are important.[7] Gubernatorial leadership also can aid in shaping the content and direction of hazardous waste policies. Governor Jerry Brown (D-California) actively championed a statute designed to encourage alternatives to the land-based containment of toxic wastes in the early 1980s, a law that predated the enactment of HSWA.[8]

Last but not least is the role played by the public in pushing for stronger hazardous waste laws. Greater activism by citizen and environmental organizations is often prompted by media attention to pollution problems which can, in turn, propel hazardous waste policy issues to the top of the governmental agenda.[9] In some cases, issue visibility of this sort can result in an occasional victory by an aroused public over a well-financed and organized coalition of business and industry groups. California voters adopted one of the strictest hazardous waste "right-to-know" laws in the United States in 1988 despite strong industry opposition. Ordinary citizens also have affected policy by thwarting the implementation of hazardous waste facility siting decisions, giving rise to the well-known acronym NIMBY (for "not-in-my-backyard").

What about policy content? It is instructive to note that state autonomy in developing RCRA-style legislation is achieved at the expense of conformity to uniform federal standards. A sizable number of jurisdictions exert the minimum degree of activity needed to obtain program control. State officials realize that EPA officials possess both oversight and preemptive management authority; hence, for some, a risk-avoidance strategy is pursued. Evaluation of this sort requires information gleaned from program audits, reporting requirements, grant reviews and awards, and enforcement actions, and these tasks can be completed more easily with similar statutory components in place. By the mid-1980s, fifteen states had adopted hazardous waste programs that essentially parroted federal statutory language.[10]

While state programs are shaped by federal influences, they are by no means a homogeneous group devoid of independent policy initiatives. Between-state differences are found on all sections of the RCRA management program. Guidelines covering the definition and classification of hazardous wastes found in most states are broader than those offered by EPA. According to Doyle, state laws typically incorporate EPA's list of hazardous wastes within their own universe of wastes to be regulated.[11] Additional criteria may be used to identify a substance as hazardous, such as carcinogenic properties (Rhode Island), acute toxicity (California, Oregon, and Minnesota), and mixtures of hazardous and nonhazardous wastes (Missouri and Oregon).[12] A smaller number of jurisdictions identify and control a given waste according to the degree of hazard it poses to public health or environmental quality. For example, the state of Washington classifies hazardous substances as "dangerous waste" or "extremely hazardous waste," and uses more stringent criteria to regulate wastes contained in the latter category.[13]

States also must develop policies affecting waste generators and transporters within the context of RCRA. The key instrument of management control is a tracking system that uses an industry manifest to monitor the whereabouts of waste from its point of generation to its ultimate disposal. The owner or operator of the tsd facility receiving a shipment of waste is responsible for returning a copy of the manifest to the generator. If the waste-generating entity does not receive the manifest within 35 days of the original shipment, the firm is responsible for contacting the transporter and the facility manager to determine what happened. Failure to receive a satisfactory accounting of the missing waste should result in action taken by company representatives to notify EPA of the discrepancy. Regulatory success in this case depends on the honesty and energy of regulated entities.

Some states are more risk-averse than others when deciding how to deal with the off-site shipment of hazardous waste. New Jersey requires that a copy of the manifest be received by state regulatory officials before any wastes can leave the premises of the generator, while other states such as Illinois and Maryland require like documentation within a very short period of time after the waste has been delivered to the tsd facility. According to state officials interviewed by Lennett and Greer, prenotification requirements are preferable to a post hoc requirement that all discrepancies be reported to the relevant agency as a policy approach because of serious administrative problems found to be associated with the filing of exception reports.[14]

RCRA also deals with the management of tsd facilities. It calls for the promulgation of environmental quality standards to ensure that facilities operate cleanly and safely and for the issuance of permits to owners/operators by EPA or state regulatory officials (see discussion of permitting in Chapter 3). Receiving the requisite authority to issue permits is particularly important for state regulatory officials. Since wastes are received at a destination site where prospective risks associated with improper management procedures are high, it is necessary for administrators to have the latitude to specify minimum conditions for a safe and financially sound facility before it becomes operational.

Most state legislatures establish guidelines for the issuance of permits pertaining to design and operating standards, financial liability provisions, and, in some cases, the use of specific management practices. For example, Arkansas requires that a facility provide evidence of financial responsibility and liability insurance at a level set by the state

Department of Pollution Control and Ecology. Other requirements include plans dealing with spill prevention, groundwater monitoring, and postclosure operations.[15] The latter concern is resolved in several states such as Arizona, Connecticut, Kentucky, Ohio, and Wisconsin by a legislative provision that combines long-term state management of defunct sites with a funding mechanism based on fees paid by tsd facility operators.[16]

States also are shouldering significant policy responsibilities in the cleanup of contaminated sites. Within the context of the federal Superfund program, states participate in cost-sharing arrangements ranging from 10 percent to 50 percent of estimated cleanup expenditures (the subnational match is proportionately greater if the site is located on state land) and also may assume lead agency status in site management decisions. Following the completion of remedial action at an NPL site, states are expected to absorb all subsequent operation and maintenance costs.

Beyond actions mandated or allowed under SARA, states also have addressed the problem of contaminated sites that are large enough to threaten public health or environmental quality but too small to qualify for NPL designation. Most states have enacted legislation that provides emergency response funds, identifies a department or agency responsible for enforcement, establishes a state priority list, and enumerates decisional criteria used to select the appropriate management option for site cleanup purposes.[17] Such programs vary considerably in terms of statutory coverage as well as level and type of funding.

Enforcement authority is often but not always lodged in bureaus or departments in charge of developing hazardous waste programs. A minority of states (11) implement state Superfund laws through other environmental protection statutes with similar or overlapping policy content such as groundwater quality. For example, Michigan carries out a program using enforcement staff responsible for an array of pollution-control policies.[18]

Priority lists are used in twenty states to better allocate scarce resources for cleanup actions. Some states pattern their list after the NPL found in the federal Superfund program. Under this approach, all sites reaching or exceeding a threshold score based on the Hazard Ranking System are identified as hazardous. This may take the form of an inventory, which allows greater latitude to state officials in determining which sites to address first, or it may rank sites according to the degree of risk and/or the amount of remediation required.[19]

Other programmatic components of state Superfund policies include provisions for citizen lawsuits (15 states), victim compensation (11 states), and the certification of property as free of hazardous waste contamination before it can be sold (five states).[20] Inclusion of citizen lawsuits allows individuals or groups who are currently or potentially affected by the release of hazardous substances to initiate a civil action against the responsible party for the purpose of requiring corrective action or the prevention of further damage. Victim compensation policies are designed to aid those affected by the release of wastes by providing funds for the replacement of water supplies or relocation.

Property transfer laws are a relatively recent policy innovation stemming from a New Jersey statute adopted in 1983. It required all owners of industrial properties to clean up on-site wastes before the business could be shut down or sold. The effectiveness of the latter policy is lauded by Doyle, who points out that one-and-a-half years' experience with the new law resulted in the expenditure of $70 million to clean up 255 sites, compared with the appropriation of $88 million for site cleanup actions after ten years of experience with the state Superfund program.[21]

REGULATORY TRENDS

The mode of regulation which best describes early hazardous waste policy initiatives is the "directive" or "command and control" approach.[22] Public officials prescribe courses of action available to regulated entities or individuals. This may take a variety of forms; a ban on the production of chemicals found to be carcinogenic is certainly the most restrictive example, but the use of goals and timetables to accelerate industry compliance with hazardous waste policies, prescriptive rulemaking, and permits also can be classified as directive. Failure on the part of regulated firms to comply with regulatory objectives can result in the imposition of sanctions such as fines and/or incarceration. The reward structure for those affected is largely negative in tone; that is, noncompliance is followed by apprehension and punishment.[23]

An alternative means of achieving regulatory goals is the incentives-based approach.[24] Rules tend to be less prescriptive, focusing more on performance objectives than design criteria. Companies are given greater latitude to design abatement strategies that are both cost-effective and environmentally beneficial. Examples of economic incentives include fee structures, taxes, changing conceptions of liability, recycling, and greater use of waste-exchange systems to encourage waste management options occurring at the production phase rather than options requiring continued use of land-based containment. Industry officials are understandably more supportive of this approach, since questions of scale and flexibility can be more easily taken into account in meeting hazardous waste policy objectives.

The imposition of fees or taxes does not restrict the range of options available to industry officials, but it does increase the expense of maintaining the managerial status quo. Waste-end taxes or fees can be levied on the amount of hazardous substances actually produced and can be reduced or eliminated if the regulated entity chooses to employ source-reduction technologies or otherwise demonstrates a decline in the volume of pollutants destined for land-based disposal.[25] An important issue is the level of revenue extraction needed to generate a given increment of waste reduction. Fees have been used to induce industries in Missouri, California, and Ohio to avoid the selection of management strategies that are less expensive but environmentally suspect.[26]

Another way of increasing the costs of less desirable waste management practices is achieved by the manipulation of liability standards. Under the Comprehensive Environmental Response, Cleanup, and Liability Act (CERCLA), the joint and several standard is used, which requires all parties contributing to the contamination of a given site to participate financially in subsequent cleanup operations. This is potentially a very powerful incentive for industries to place less reliance on off-site disposal options, but it cannot be used in states that depend on non-"Superfund" enforcement authority. If state cleanup orders are based on RCRA-type laws, action taken against a particular party requires proof that RCRA was violated at the time waste disposal actually occurred.[27] Without the necessary statutory authority, the burden of proof is shifted from the potentially responsible party (PRP) to the enforcement agency, and the usefulness of liability standards as an economic incentive is greatly attenuated or lost.

Insurance requirements mandated by SARA for owners of underground storage tanks containing petroleum or other hazardous chemicals represent another form of liability standard that can promote desired "front-end" waste management practices. Companies must internalize the costs of leak prevention/detection in order to obtain insurance coverage required before the start-up or resumption of business operations. Other marketplace actors are, in effect, supplementing directives issued by state regulatory agencies to ensure compliance with hazardous waste regulations. At present, a major drawback to the effectiveness of this approach is the reluctance of major insurance companies to offer coverage for prospective environmental damages. Some states have responded to the challenge by forming risk pools to assist owners of underground tanks or tsd facilities. States would become insurers of last resort under this plan.[28]

Economic incentives also are used to encourage greater industry use of waste-disposal strategies other than land-based containment. Most public officials are convinced that producing fewer wastes, altering the chemical composition of hazardous substances to reduce their toxicity, or recycling would be preferable to off-site disposal if cost were no object. However, the political feasibility of assorted management options is clearly affected by the interplay of economic and technological factors. Two of the more commonly mentioned approaches here are governmental financial assistance to pollution-generating firms and waste-exchange systems.

Industry officials favor governmental efforts to obtain easier access to the capital needed to undertake changes in facility design or process substitution. One example is the development of special loan programs for waste-disposal firms specializing in innovative treatment or disposal technologies. Complementary strategies discussed by the Office of Technology Assessment include stretching out the allowable time span for repayment and delegating sufficient financial management authority to state regulators to guarantee private-sector loans or to issue tax-free bonds to finance these loans.[29]

Waste-exchange systems represent yet another incentives-based approach, which is used in conjunction with recycling. One type of exchange system is limited to the role of clearinghouse: matching companies that have hazardous wastes to sell with prospective buyers. An integral part of the process is the guarantee of confidentiality by waste-exchange officials to ensure that trade secrets are not jeopardized. A broader set of responsibilities is shouldered by a waste-materials exchange, which combines the information brokerage role with the physical transfer of the product itself. This requires the presence of staff and equipment so that the chemical properties of the waste can be verified. Not surprisingly, overhead and service costs are higher with the latter approach. In general, the opportunities for increasing the volume of recycled wastes through the use of waste-exchange systems are limited by technical factors. The percentage of hazardous wastes with any sort of market potential is quite small. Nevertheless, a few exchanges are operating in California, Iowa, and the New England states.[30]

Some states are less receptive than others to the supplemental use of economic incentives. Why is this the case? Public officials are likely to be influenced by several factors, including administrative resistance to change, media attention to health and environmental risks posed by exposure to hazardous wastes, and the presence or absence of assorted policy characteristics at the state level.

An especially plausible reason for maintaining a directive approach to regulation lies in the nature of bureaucratic inertia. Organizational routines or standard operating procedures are developed to carry out program management responsibilities, and administrators are reluctant to deviate from established practices. Over time, an administrator's professional reputation becomes increasingly intertwined with the justification for and subsequent defense of existing regulations before legislators and clientele groups. Thus, the consideration of alternative regulatory forms collides with organizational "sunk costs." A status quo orientation also may be reinforced by budgetary constraints or by a lack of turnover within the initial cadre of administrators.[31]

The feasibility of incorporating economic incentives within state regulatory structures also is affected by the visibility of hazardous waste problems. Media attention to the discovery of abandoned dump sites or the accidental release of hazardous substances into the environment often is accompanied by public pressure to strengthen hazardous waste regulations. Attempts to develop an incentives-based approach under these circumstances would likely be characterized as governmental capitulation to corporate polluters or an unwarranted effort to measure the worth of human lives in economic terms.

State policy characteristics also can affect policymakers' response to differing modes of regulation. One study has concluded that states ranking high on policy leadership and fiscal health indices are more likely to incorporate economic incentives.[32] This finding is compatible with the socialization of cost thesis mentioned earlier as well as the notion that some states have been looked to by public officials in other states as a particularly useful source of ideas for the development of regulatory policy initiatives. California, Minnesota, New Jersey and Illinois are states that have historically demonstrated leadership in experimenting with incentives-based approaches to regulation.[33]

SITING HAZARDOUS WASTE FACILITIES

Perhaps the most controversial aspect of hazardous waste policymaking is the siting of tsd facilities. Under RCRA, Congress decided to delegate to the states decisions covering facility placement, the criteria to be incorporated in making such choices, and the scope of citizen participation arrangements. The ostensible rationale (other than avoiding a political hot potato) was that programs could be more easily tailored to meet state needs. EPA also was prohibited by law from acquiring land that could subsequently be leased to firms specializing in the disposal or containment of hazardous wastes or from building or operating a facility. State governments also possessed the power of eminent domain and other land-use powers necessary for acquiring or approving prospective sites.[34]

More than half of the states have enacted hazardous waste facility laws since 1980. There is considerable variation in statutory content, ranging from highly centralized state-centered approaches to the retention of local autonomy. Several states (nine) have opted for policies that effectively preempt local veto authority. Lawmakers in Arizona and Georgia adopted programs that greatly restrict citizen participation in the decision-making process.[35] Experts, rather than members of the public, are viewed as most qualified to pick appropriate sites for facility construction in an efficient and dispassionate manner. A governmental agency is given the responsibility for both site selection and authorization of use for facility operations.[36]

A more common policy approach identified by Andrews is referred to as state override.[37] Here state boards are given the authority to effectively block or overturn local initiatives designed to thwart hazardous waste facility development. As with preemptive siting policies, an underlying assumption of override programs is the necessity of substituting a more objective and comprehensive state perspective on waste-disposal capacity for the more parochial and emotion-laden concerns associated with NIMBYism in or near communities known to be on the "short list" of possible sites. One example of a state that uses overrides is Pennsylvania, which empowers a board lacking local representation to make site-selection decisions. Other states such as California and North Carolina use a somewhat less onerous variant that allows boards to entertain appeals from local governments chosen to house a tsd facility.[38]

At the other end of the policy spectrum are statutes that vest control over hazardous waste facility decisions in the hands of local officials. Eight states have adopted a local veto model, which is justified by arguments such as responsiveness to affected constituencies as well as the rather practical notion that siting decisions are not going to be implemented without community acceptance.[39] For example, Colorado has rejected legislative proposals calling for state preemption or override provisions because of opposition from the county commissioners. Other states are placed within this category because state officials have neglected to address siting questions at all; hence, local control becomes the de facto default option.

A fourth approach also maintains the locus of decisionmaking at the local level of government but differs from the local veto option because of procedural restraints imposed by state officials. Such limitations are undertaken to ensure that applications for the construction and operation of hazardous waste facilities are handled fairly.[40] An example is a requirement that "good-faith" siting negotiations be conducted between prospective developers and community representatives. Wisconsin has a provision of this sort—it allows state officials to penalize local governments that essentially go through the motions of participating in a policy dialogue without demonstrating any real intention of giving would-be facility operators an opportunity to convince community negotiators that waste-disposal operations can be conducted safely.

Another state operating within this framework is Illinois, which treats local approval of facility siting applications as the critical first step in obtaining a state permit. Thereafter, the burden of proof is placed firmly on project opponents. Local approval decisions can be appealed to the state pollution-control board. However, the only chance for obtaining a reversal depends on the presentation of evidence showing that the prospective site will produce unacceptable consequences stemming from incompatible land uses or a failure to consider public safety in designing the facility.[41]

Thus far, there is little evidence to suggest that any of these policy approaches is more effective in siting tsd facilities. In a study of facility siting efforts initiated by private-sector firms between 1980 and 1986, no significant differences were found in the success rate of states pursuing alternative policy paths.[42] While we acknowledge the possibility that in some states policies have not been in place long enough to evaluate their effectiveness, it is instructive to note that other students of hazardous waste facility siting place more emphasis on policy process than on the assignment of a greater or lesser degree of formal authority to state government. Of particular interest here is research focusing on public perception of risks associated with physical proximity to prospective sites and decision-making procedures that are aimed at the amelioration of safety-related fears linked to the development of NIMBY sentiment.[43]

One avenue of inquiry revolves around the manipulation of incentives, such as public compensation schemes, the increased dissemination of information to community residents about safety issues prior to the site-selection stage, and the formal representation of citizens in the decision-making process.[44] For example, O'Hare has examined the potential usefulness of monetary compensation as a means of mitigating the social, psychological, and economic impacts of a siting decision on the host community. Compensatory schemes such as grants or contingency funds are designed to redress the imbalance between the benefits received by citizens dispersed throughout the state and region from safer waste-disposal practices and the costs borne by the local jurisdiction as the result of facility operations.[45]

Whether the public will respond favorably to incentives of this sort is questionable. Results from a pair of survey-based studies indicate that offering a financial settlement as a consolation prize does not appreciably affect the willingness of community residents to house a toxic waste facility.[46] At issue for many people is the specter of companies attempting to get public acquiescence by buying them off. Or, as Pitney has noted, "bile barrel burdens do not readily translate into dollars and cents."[47]

Other noneconomic approaches, such as public education or citizen participation in siting negotiations, have shown more promise as ways of reducing fear associated with tsd facility development.[48] Studies focusing on the attitudes of community residents in Massachusetts and Wyoming toward an array of compensatory measures revealed a preference for policies requiring the dissemination of information to the public prior to the site-selection phase or for risk-mitigation proposals, such as more frequent tests of groundwater quality.[49] Residents are concerned about not only safety issues but also acquiring or maintaining greater control over site-selection decisions.[50]

While the mix of incentives to be used represents a crucial step in persuading some individuals or groups to accept a toxic waste facility, it is also apparent that this task can be achieved in some communities more easily than others. One historic pattern of siting decisions initiated by industry officials that has increasingly drawn criticism on equity grounds is the practice of targeting less-populated rural areas as a desirable location for waste-disposal activities. From an industry perspective, the advantages are obvious. These sites are distant from more populous urban areas, thereby minimizing public-health risks; they also are less likely to provoke community opposition. However, the argument raised against the continuation of "out of sight, out of mind" siting practices is that communities with a higher percentage of residents who are poor, minority, and uneducated are asked to house a disproportionate share of the pollutants while the jobs and economic benefits created by waste-generating firms remain elsewhere.[51]

Other factors also are at work. One influential study focused on the likelihood of hazardous waste facility approval in relation to the political and demographic characteristics of affected local jurisdictions in New Jersey. Some of the more important factors included the economic impact of a prospective facility on a community, its compatibility with existing land-use patterns, and residents' perceptions of costs and benefits.

Local support was obtained in one case after managers of a waste-generating firm already operating within city limits demonstrated need for additional waste-disposal capacity. The proposed site was not located in close proximity to populated areas. Failure to approve the request clearly would have produced an economic setback. Thus, need was tempered by assurances of safe facility-management practices.[52]

The politics of facility siting decisions also is affected by the attitudes and actions of public officials. For many, the most difficult task is reconciling the need for technical expertise in making site-selection decisions with the potentially unsettling effects of public participation. One study found that state hazardous waste administrators and chemical industry officials alike felt uncomfortable with the notion of allowing citizens to take part in decisions unless their involvement was restricted to testifying at other public hearings or serving as representatives on siting boards. Respondents were strongly opposed to more active modes of public participation, such as requiring local approval for siting proposals in a referendum election or encouraging the debate of these issues in local political campaigns.[53]

Experimentation with institutions designed to resolve policy gridlock of this sort has taken place in California. A sizable and growing disparity between the production of hazardous waste and the availability of off-site disposal capacity led to the realization by virtually every local official and political organization within the Los Angeles–San Diego corridor that new tsd facilities were needed. Most recognized that an expressed willingness on the part of any city or county council member or state representative to consider housing a facility in his or her district would be tantamount to political suicide. On the other hand, there was substantial agreement on the part of disparate political interests that a preemptive siting decision made by state officials in Sacramento was to be avoided at all costs.

What could be done to handle this dilemma? Following the suggestions of an enterprising local official, support was developed through an exhaustive series of discussions held between representatives of city and county governments, environmentalists, industry leaders, public-health organizations, academics, good-government groups, planners, and technical experts for the creation of a regional public authority to address the hazardous waste facility siting issue. A key objective of these deliberations was to obtain a decision-making structure that was sensitive to industry and environmental concerns alike and could be trusted to offer policy recommendations that would be perceived as fair by the participants and the general public.

The product of these negotiations was the Southern California Hazardous Waste Management Authority (SCHWMA). In the short run, a key achievement was getting key policy participants to acknowledge the legitimacy of contending points of view. For environmentalists, this meant recognition that a sound waste management strategy had to be discussed within the context of minimizing rather than eliminating environmental risks. Industry officials were concerned about the protection of trade secrets but agreed that it was necessary to share and publicize information about waste-disposal technologies.

Each party recognized that progress could only be achieved by focusing on problems instead of adversaries. The authors of a case study suggested that SCHWMA did succeed in creating a consensus-based management style oriented toward making—and implementing—tough decisions in a contentious political atmosphere.[54] While success in terms of siting new facilities has not occurred (as of 1992), there are currently three facilities under review by SCHWMA that may yet produce the desired results.

In short, siting decisions continue to be constrained by NIMBY-related factors. Some contend that this trend has produced more good than harm, since industries are now beginning to consider waste-reduction strategies more seriously as a consequence of mounting political and economic costs associated with off-site disposal practices.[55] Regulatory costs, in other words, are increasingly borne by waste-generating firms in accordance with the "polluter pays" principle.

Others argue that hazardous waste tsd facilities will be necessary even if firms strive to decrease the amount of waste produced. Thus, attention is directed toward procedural solutions such as increasing the role of the public in both site selection and the oversight of facility operations, and political solutions, such as requiring substate involvement in the preparation and submission of a siting plan to state authorities to be used whenever additional waste disposal capacity is needed.[56]

LOCAL POLICY INITIATIVES

When Herson revealed the scholarly neglect of local politics several decades ago, he coined the phrase "the lost world of municipal government."[57] The term applies with nearly equal force to the area of local environmental politics, including hazardous waste. Nevertheless, city and county officials have become increasingly concerned about waste-disposal problems, partly because of actions mandated by federal and state statutes and partly because of the responsibility they bear for protecting the health and safety of their constituents.

The most visible—and controversial—aspect of hazardous waste policy affecting local government is Section 3007 of RCRA, which delegates the responsibility for siting tsd facilities to state government. State officials clearly have the upper hand both legally and politically and have in some cases adopted highly restrictive statutes that inhibit efforts by city and county officials to halt the construction of waste-disposal facilities (see preceding section). Other states, notably Florida and California, have created a more active role for local government. Counties in both states must become involved in bona fide planning efforts to identify prospective sites for the storage or disposal of wastes. Initial cooptation of local officials in this fashion may dim the expression of NIMBY-related sentiments of community residents in subsequent site-selection decisions.

Program requirements for the protection of groundwater quality represent another major concern for local officials, since U.S. residents depend on the withdrawal of subsurface water resources for about half their water supply.[58] An emerging threat is posed by the growing number of leaking underground storage tanks containing petroleum or other hazardous chemical substances. Or, as Doyle has indicated, "of the 1.4 million gasoline tanks nationally, some experts estimate that between 75,000 and 100,000 are leaking and perhaps up to 350,000 will begin leaking in the next five years."[59] Congress responded by incorporating a key provision within HSWA in 1984 that placed these tanks under EPA's regulatory umbrella.

While EPA and state-approved agencies are ultimately accountable for program performance, the actual task of identifying problem tanks, installing leak-detection devices, and monitoring them is often carried out by local fire departments, or, in some cases, by public-health agencies. An advantage of ceding program control to a public safety organization lies in the efficiency gain for emergency response actions following the accidental release of hazardous chemicals into the ground. Priority is given to ameliorating immediate threats to public safety.

A possible disadvantage lies in ignoring migratory contaminants that may jeopardize groundwater quality if cleanup actions do not accompany the repair of leaks or tank removal. In some jurisdictions such as Larimer County, Colorado, an interlocal agreement between the Health Department, the City of Fort Collins, and the Poudre Fire Authority calls for the sharing of information coupled with a functional division of responsibilities. This ensures that multiple policy objectives can be addressed, including the protection of water supplies and environmental quality as well as public safety.

Yet another federal policy initiative requiring substate participation is the Community Right to Know Act of 1986 (Title III of SARA), which provides local residents with access to information about the amount, type, and location of hazardous waste currently stored in or near their community. Thus, any individual or company possessing a given quantity of one or more hazardous chemical substances "must notify the state, the local emergency planning committee and the local fire department, telling them where the material is located, how much is usually on hand and what health risks it poses."[60]

In theory, this law can provide a much-needed jump start to the preparation and, if necessary, the implementation of an emergency preparedness plan dealing with the accidental release of toxic chemical substances into the environment. Local officials coordinate the dissemination of information to the public and the evacuation of nearby residents. The key actor is typically the local emergency planning committee (LEPC), which includes governmental officials and representatives from industry, the media, and the public. Whether it works as intended largely depends on the attitude projected by local leaders. A task of this sort can be depicted as an opportunity to develop an effective response to emergencies such as chemical spills or as another unwanted mandate imposed by the federal government with attendant reporting requirements but no funding.

The experience of Columbus, Ohio, with "right-to-know" legislation is illustrative. Following the enactment of Title III by Congress in 1986, the Columbus Health Department took the lead in organizing the Chemical Emergency Preparedness Council (CEPAC), which consisted of community leaders, health officials, industry management and active citizens. This effectively satisfied the legislative requirement for the LEPC to hasten the implementation of the federal law.[61]

In addition to following the federally mandated reporting requirements for dangerous chemicals, CEPAC went a step further in initiating a "volunteer program" for companies possessing hazardous chemical substances in quantities below the minimum threshold levels. The city of Columbus also carried out extensive computer mapping of sites known to be handling or storage facilities for these wastes and their proximity to significant landmarks such as schools, parks, or hospitals. Finally, a public-education program was undertaken to inform community residents about the location of sites containing hazardous materials, the degree of risk to the public posed by each, and the evacuation plan to be used should an emergency arise.[62]

What have local governments done on their own to address hazardous waste problems without prompting from federal or state mandates? In a national survey of city and county officials, Scheberle and Lancaster found that a combination of approaches were used. Less formally, community leaders were involved in educational programs designed to inform the public about safe waste-disposal practices. More legalistic approaches included the creative use of land-use planning and the adoption of ordinances dealing with the collection and disposal of household hazardous waste, the transport of waste in or near a community, and the exclusion of wastes from municipal landfills or wastewater treatment plants.[63]

Some communities also have attempted to shift the burden of enforcing hazardous waste laws to the private sector. Dade County, Florida (the greater Miami area), adopted an ordinance forcing property owners to confront problems of contaminated land by requiring sellers of property to present a certificate of compliance before a title could be transferred—a policy closely patterned after the New Jersey statute discussed in a previous section. A similar law requiring a site history and, if necessary, the cleanup of pollutants prior to the issuance of construction permits was enacted by public officials in San Francisco.[64] Thus, police powers can be combined with economic incentives to produce environmental protection programs that can be implemented with fewer public-sector financial and administrative resources.

ALLOCATION OF RESOURCES

Effective policy implementation clearly requires sufficient levels of funding and staff. Fiscal resources are drawn from disparate revenue pots, but the federal and state treasuries are the largest contributors. Federal payments to state-approved agencies take the form of categorical grants for the management of RCRA-type programs (a minimum state match of 25 percent is required) and the transfer of monies from trust funds established for the cleanup of contaminated sites on the NPL and leaking underground storage tanks containing petroleum or other hazardous chemical substances. Funding also is obtained from state legislative appropriations and other sources.

EPA grants have consistently provided a sizable portion of state financing for hazardous waste programs. A report issued by the Senate Environment and Public Works Committee in 1983 indicated that states were quite dependent on Washington's monetary pipeline to cover operating costs. Some states, such as Kentucky and Montana, derived over 80 percent of their RCRA funding from external sources, while others, such as California, shouldered nearly 90 percent of program costs from own-source revenues.[65] Moreover, despite widely publicized efforts by OMB Director David Stockman to reduce or even "zero-out" pollution-control grants to the states, relatively few jurisdictions at this level acted to replace federal cutbacks with an infusion of monies from their own coffers.[66]

Hazardous waste management grants to the states have steadily increased since the mid-1980s. While states received $46.7 million from the federal government in 1985, this figure rose to $68 million by 1990.[67] Unfortunately, there are no data detailing how these funds are spent and relatively little information about actual state expenditures for RCRA. Survey results from a study conducted by the Association of State and Territorial Solid Waste Management Officials in 1987 concluded that the average state contribution to overall program costs in percentage terms was approximately 40 percent.[68]

Data obtained from a 1988 survey of state hazardous waste managers revealed an upward trend in appropriations for the state share of RCRA expenses. Over three-fourths of those responding (N = 39) reported that their program budget had increased over the past five years, while a smaller percentage revealed that the size of the budget had decreased (12 percent) or remained the same (9 percent).[69]

State Superfund spending varies considerably in terms of size, funding sources and conditions of use. Six states employ more than 100 staff in activities related to the cleanup of contaminated sites, with New Jersey's 600-plus employees leading the way. All of these are states with a large number of sites on EPA's NPL as well as a sizable number of smaller tracts of polluted land. Many states (25) employ between 11 and 50 personnel, while 15 states have ten or fewer staff with Superfund-related responsibilities.[70] Obtaining comparable figures is a difficult task, since some states use staffers for more than one pollution-control program and in other states there is a wide gap between the number of authorized positions and the number of positions that are actually funded.

What about funding sources for state Superfund programs? Monies are typically obtained from a combination of federal grants and state general and/or cleanup funds although four states are reliant on Washington as the sole source of financial support for their cleanup programs. Federal CERCLA grants are available to augment state cleanup activities including oversight actions, site-lead responsibilities, and database maintenance, while state funds can be received from numerous sources including legislative appropriations, bonds, waste-handling fees, cost recoveries, taxes, penalties or fines, interest on fund monies, and transfers from other funds or accounts. The most important sources of revenue are appropriations and fees (19 states each), followed by bonds (12 states), penalties or fines (11 states), and taxes (nine states).[71]

Finally, it is important to note that funds may be applied to a wide array of activities. Expenditures usually can be quickly approved for standard actions such as emergency response, remedial actions, design work, and maintenance. However, several states have expanded the range of permissible uses to include victim compensation, development of new hazardous waste facilities, or grants to municipalities and local governments.[72]

ENFORCEMENT

The enforcement of RCRA represents a particularly daunting challenge for state and local administrators. The process of identifying noncompliant parties has become more difficult since the removal of the small-quantity-generator exemption in 1984. A consequence is the increased complexity of enforcement as a policy problem. The key question involves the application of sanctions to a large and diverse group of violators in an evenhanded manner.[73] Administrative effectiveness may well be pursued at the expense of consistency and fairness. Program managers are placed in the position of targeting larger firms, thus ensuring maximum public protection for the investment of limited resources or implementing the law with little or no regard for the size of the regulated entity.

In a larger context, the notion of jurisdictional commitment to hazardous waste policy goals is examined here by comparing the use of enforcement actions across the states. In addition, other state attributes that can potentially contribute to a better understanding of between-state differences in regulatory effort are analyzed. Of particular importance are economic factors, such as the degree of dependency on chemical manufacturing or processing as a source of jobs for state residents, and political factors, such as the distribution of partisan or ideological orientations.

It is necessary to acknowledge that variable use of administrative tools by state regulators may reflect, in part, a preference for a different style of enforcement. However, the analysis of cross-state differences does provide one means of determining whether states in general are up to the task of administering waste management programs or whether a residual fear of economic blackmail from pollution-generating companies has adversely affected regulatory efforts in states that are more economically dependent on the production, distribution, or disposal of toxic chemical wastes. Such findings carry important policy implications for the intergovernmental assignment of policymaking responsibilities.

Let us turn to a discussion of RCRA enforcement efforts taken by state officials. At the outset, the pace of program implementation ranged from zero to glacial. Since most of the major rules affecting the hazardous waste management program were not in place before 1981, states did not begin to enforce the law in earnest until well into President Reagan's first term of office. Delegated authority to implement RCRA also was slowed by resource limitations (see Chapter 4).

The effects of these decisions are reflected in the number of actions taken. State officials conducted a total of 19 inspections nationwide in 1981, increasing to 93 in 1982, 438 in 1983, and over 5,000 in 1984. A similar pattern is found for the issuance of both warning letters and administrative orders, a not unexpected finding since enforcement efforts depend on information derived from inspections. Nor is there any evidence to suggest that legal actions took precedence over regulatory efforts. Few cases were referred to state attorneys general for disposition in the courts.[74]

Since 1984, both inspections and enforcement actions at the state level have risen steadily. The overall number of inspections peaked at just over 10,000 in 1986 but has remained fairly stable since then with better than 9,000 tsd facility visitations in 1987 and 1988. Again, similar results are obtained for administrative actions such as warning letters and administrative orders. Litigation remains a little-used enforcement tool, although the number of referrals increased considerably from 1988 to 1989.[75] One possible explanation for this rise is the greater emphasis placed on the implementation of hazardous waste programs by EPA Administrator Reilly, a course of action that may have stimulated similar efforts at the state level.

What about between-state differences in enforcement efforts? As we noted in a previous section, some analysts contend that regulatory commitment on the part of state officials is inversely related to the economic importance of pollution-generating firms within the state. Because states compete for these industries, policymakers are reluctant to strictly enforce regulatory policies that would put resident firms at a competitive disadvantage with similar companies in neighboring states.[76] Others suggest that enforcement problems can be attributed to political rather than economic factors.[77] Partisan orientations and a willingness on the part of state policymakers to adopt tough laws are two prominent indicators of regulatory effort.

Regional differences in the enforcement of RCRA by state officials are quite evident.[78] An analysis of regulatory data collected in 1987 indicates that southern states are much more likely to be inspected than states located elsewhere, a finding that is partially attributable to the severity of that region's waste management problems (the states of Texas, Tennessee, Louisiana, and Alabama collectively account for more than a quarter of all hazardous wastes generated in the United States). The use of warning letters as a means of inducing compliance with environmental regulations is most pronounced in the Midwest, while the issuance of administrative orders is somewhat more likely to occur in the southern and northeastern states.

On the other hand, enforcement actions of any sort are least likely to be taken in the West, a tendency that is especially noticeable if California is excluded from the analysis. This area actually contains two distinct subregions. The Pacific Coast tandem of California and Washington more closely resembles the heavily industrialized states of the Midwest and the Northeast, with their attendant problems of waste disposal. These states contrast sharply with states located in the Rocky Mountain region, which are less dependent on an industrial economic base than a combination of resource extraction (such as timber, minerals, and energy production) and tourism.[79]

How do economic and political factors affect interstate differences in state enforcement decisions? The data indicate that the severity of waste management problems within jurisdictional boundaries is the main catalyst for agency action. Inspections, NOVs, and administrative orders are carried out more frequently in states that produce or store large quantities of hazardous wastes. This is consistent with prior research that established a link between problem magnitude and the strength of state hazardous waste policies, even after other sources of influence were taken into account.[80]

A number of other political and economic factors contribute to our understanding of state enforcement decisions as well. The comprehensiveness of state hazardous waste laws and the degree of political control by the Democratic party are moderately related to the frequency of both inspections and the issuance of administrative orders. While these findings are not unexpected, we were surprised to discover a positive statistical relationship between economic dependency on the production or distribution of chemical products and policy enforcement decisions. This runs counter to the notion that state regulatory actions are driven to a greater extent by possible economic repercussions such as industry relocation than by the perceived seriousness of waste management problems.

A less visible but important area of enforcement involves the monitoring of small businesses subject to hazardous waste regulations since the enactment of HSWA in 1984. Unlike tsd facilities, many smaller firms, such as auto repair shops or dry cleaners, not only are producing waste but are doing so in close proximity to residential neighborhoods. In addition, there is reason to believe that noncompliance rates for these small-quantity generators (SQGs) are quite high. A study of SQG waste management practices concluded that illegal disposal activities are widespread. This conclusion resulted from reviewing multiple sources of information, including interviews with representatives of commercial waste management firms and with state hazardous waste program managers (data showing that a sizable number of SQGs nationwide have not acquired an EPA identification number) and other published accounts of enforcement problems in local government.[81]

Why have SQGs become such a major source of irritation for local officials? Perhaps the key factor is the lack of deterrent effect associated with HSWA. The costs of compliance are often high, particularly for SQGs operating on the margins of profitability, while the risks of getting caught illegally disposing of hazardous waste and being punished for such actions are relatively slight. State or local resources devoted to policing noncompliance among small business operators are usually small to nonexistent. And even well-meaning business owners wishing to comply with the law find that off-site disposal costs are financially prohibitive because potentially useful options such as transfer stations or transportable treatment units are not available. Finally, a significant number of SQG operators do not use effective management practices because they are unaware of HSWA and what it requires.[82]

CONCLUSIONS

An increasingly important question affecting intergovernmental regulatory programs dealing with hazardous waste or other environmental problems is whether state policymakers have the institutional capabilities or the will to manage them. How do states "measure up"? A focus on policies reveals that the minimal requirements for primacy have been met in most states, but it is equally evident that some jurisdictions do vastly more in terms of program development than others.

On the positive side, experimentation occurs on a variety of fronts, such as the New Jersey law requiring proof that land and/or buildings are not contaminated before property can be sold, or the enactment of statutes in California and Florida calling for the preparation of county plans for siting a tsd facility that are then forwarded to state authorities. Local government officials also have begun to consider regulatory initiatives to inform and protect their constituents. However, most states offer the basic statutory package and little more.

Progress on the financing and enforcement of hazardous waste policies has occurred since the mid-1980s. This is in part a response to changes in the commitment to program enforcement expressed by EPA officials. More money was budgeted at the federal level for the enforcement of RCRA and Superfund. Congress also has authorized increases in grant monies to supplement state program operations. Generally speaking, state budgets rose between 1984 and 1988.

Similar shifts have occurred in the field as well. Overall, both enforcement actions and the referral of civil and criminal cases to state attorneys general are on the rise. However, important between-state differences remain in the number and variety of administrative activities, findings that are largely attributable to the relative severity of pollution problems. At the local level, a particularly glaring deficiency in terms of funding and enforcement is the current lack of attention given to small-quantity generators.

The critical question is whether these increases in funding and staff for enforcement purposes are sufficient, taking the magnitude of waste-disposal problems into account. Are these largely stopgap measures that are serving to keep existing levels of pollution from getting worse, or can a modicum of progress be identified in some areas that offers hope for future improvements in environmental quality along with fewer public-health risks? A fuller evaluation of hazardous waste policymaking and suggestions for policy and administrative change are offered in the concluding chapter.

NOTES

1. James P. Lester, James Franke, Ann O'M. Bowman, and Kenneth Kramer, "Hazardous Wastes, Politics and Public Policy: A Comparative State Analysis," *Western Political Quarterly* (June 1983).

2. Ibid.

3. This point was initially made in Malcolm Getz and Benjamin Walter, "Environmental Policy and Competitive Structure: Implications of the Hazardous Waste

Management Program," *Policy Studies Journal* (Winter 1980). Subsequent efforts to critically assess the notion that environmental regulation has become secondary to economic development goals include David Goetze and C.K. Rowland, "Explaining Hazardous Waste Regulation at the State Level," *Policy Studies Journal* (September 1985); C.K. Rowland, S.C. Lee, and David Goetze, "Longitudinal and Catastrophic Models of State Hazardous Waste Regulation." In Charles Davis and James P. Lester, eds., *Dimensions of Hazardous Politics and Policy* (Westport, Conn.: Greenwood Press, 1988); Richard Feiock and C.K. Rowland, "Environmental Regulation and Economic Development: The Movement of Chemical Production Among States," *Western Political Quarterly* (September 1990); and Richard Feiock and Charles Davis, "Can State Hazardous Waste Regulation Be Reconciled with Economic Development Policies? An Empirical Assessment of the Goetze-Rowland Model," *American Politics Quarterly* (July 1991).

4. Consult John P. Blair and Robert Premus, "Major Factors in Industrial Location: A Review," *Economic Development Quarterly* (February 1987); and Charles Davis, "State Environmental Regulation and Economic Development: Are They Compatible?" *Policy Studies Review* (Winter 1991).

5. Bruce Williams and Albert Matheny, "Testing Theories of Social Regulation: Hazardous Waste Regulation in the American States," *Journal of Politics* (May 1984).

6. Ibid.

7. Ibid.

8. David Morell, "Technological Policies and Hazardous Waste Politics in California." In James P. Lester and Ann O'M. Bowman, eds., *The Politics of Hazardous Waste Management* (Durham, N.C.: Duke University Press, 1983).

9. Adeline Levine, *Love Canal: Science, Politics and People* (Lexington, Mass.: Lexington Books, 1982). See also Riley Dunlap, "Public Opinion and Environmental Policy." In James P. Lester, ed., *Environmental Politics and Policy* (Durham, N.C.: Duke University Press, 1989).

10. Ann O'M. Bowman, "Hazardous Waste Management: An Emerging Policy Area within an Emerging Federalism," *Publius* (Winter 1985).

11. Paul Doyle, *Hazardous Waste Management: An Update* (Denver, Colo.: National Conference of State Legislatures, 1975) p. 10.

12. David J. Lennett and Linda Greer, *State Regulation of Hazardous Waste: A Summary and Analysis* (Washington, D.C.: Environmental Defense Fund, 1985), pp. 11–24.

13. Doyle, pp. 11–12.

14. Lennett and Greer, p. 85.

15. Doyle, p. 20.

16. Ibid., p. 20.

17. U.S. Environmental Protection Agency, *An Analysis of State Superfund Programs* (Washington, D.C.: Government Printing Office, September 1989), pp. 6–7 (hereafter referred to as EPA report).

18. Ibid., p.6.

19. Ibid., p.7.

20. Ibid., p.7.

21. Doyle, p. 76.

22. Much of this discussion is drawn from Charles Davis, "Approaches to the Regulation of Hazardous Wastes," *Environmental Law, 18* (Fall 1988). For a useful analysis of the comparative advantages of directive- versus incentives-based approaches to regulation, see Barry Mitnick, *The Political Economy of Regulation* (New York: Columbia University Press, 1980).

23. Early emphasis on directive regulation in both RCRA and Superfund is demonstrated by the establishment of goals and timetables, stringent permit requirements, and the adoption of prescriptive rules. In part, this regulatory emphasis can be attributed to the understandable desire of federal officials to borrow policy ideas from prior pollution-

control programs, notably the Clean Water Act. But it also reflects the long-standing opposition to economic incentives voiced by key political decision makers such as Senator Ed Muskie (D–Maine), who served as Chair of the Public Works Subcommittee on Air and Water Pollution.

24. The literature on this topic is voluminous. Two especially useful pieces include Thomas Ingersoll and Bradley Brockbank, "The Role of Economic Incentives in Environmental Policy." In Sheldon Kamieniecki, Robert O'Brien, and Michael Clarke, eds., *Controversies in Environmental Policy* (Albany: SUNY Press, 1986); and Stephen Elkin and Brian Cook, "The Public Life of Economic Incentives," *Policy Studies Journal* (June 1985).

25. Charles Haas, "Incentive Options for Hazardous Waste Management," *Journal of Environmental Systems, 14*, No. 4 (1984–85).

26. Robert Finlayson, "Hazardous Waste Fee Systems," *Solid Wastes Management* (September 1982).

27. EPA Report, p.27.

28. Doyle, p. 75.

29. Office of Technology Assessment, *Technologies and Management Strategies for Hazardous Control* (Washington, D.C.: Government Printing Office, 1983), pp. 30–32.

30. Katherine Durso-Hughes and James Lewis, "Recycling Hazardous Waste," *Environment* (March 1982).

31. These points are described in Anthony Downs, *Inside Bureaucracy* (Boston: Little, Brown, 1967).

32. Ann O'M. Bowman and James P. Lester, "Hazardous Waste Management: State Government Activity or Passivity?" *State and Local Government Review* (Winter 1985).

33. These states emerge as innovators most frequently from reports on hazardous waste policy developments authored by Doyle (from the National Conference of State Legislatures), Lennett and Greer (from the Environmental Defense Fund), and the OTA.

34. David Morell and Christopher Magorian, *Siting Hazardous Waste Facilities: Local Opposition and the Myth of Preemption* (Cambridge, Mass.: Ballinger, 1982). See also Michael O'Hare, Lawrence Bacow, and Debra Sanderson, *Facility Siting* (New York: Van Nostrand, 1983).

35. Richard N.L. Andrews, "Hazardous Waste Facility Siting: State Approaches." In Charles Davis and James P. Lester, eds., *Hazardous Waste Politics and Policy* (Westport, Conn.: Greenwood Press, 1988).

36. Ibid.

37. Ibid.

38. Ibid.

39. Ibid.

40. Ibid.

41. Ibid.

42. Ibid.

43. There is a sizable amount of work on the topic of risk perception and communication. One of the more influential studies is Baruch Fischhoff, et. al., *Acceptable Risk* (New York: Cambridge University Press, 1981).

44. Kent Portney, *Siting Hazardous Waste Treatment Facilities* (Westport, Conn.: Auburn House, 1991). A useful example of an effort to utilize various incentives in actual facility siting decisions is discussed in Minnesota Waste Management Board, *Charting a Course: Public Participation in the Siting of Hazardous Waste Facilities* (Crystal, Minn.: Waste Management Board, 1981).

45. Michael O'Hare, "Not on My Block You Don't: Facility Siting and the Importance of Compensation," *Public Policy* (Fall 1977).

46. Kent Portney, "Allaying the NIMBY Syndrome: The Potential for Compensation in Hazardous Waste Treatment Facility Siting," *Policy Studies Journal* (September 1985);

and Charles Davis, "Public Involvement in Hazardous Waste Siting Decisions," *Polity* (Winter 1986).

47. John J. Pitney, Jr., "Bile Barrel Politics: Siting Unwanted Facilities," *Journal of Policy Analysis and Management* (Spring 1984), p. 447.

48. See Morell and Magorian, *Siting Hazardous Waste Facilities*, and Minnesota Waste Management Board, *Charting a Course*.

49. Portney, pp. 81–89; and Davis, pp. 296–304.

50. Ibid. See also Michael Elliot, "Improving Community Acceptance of Hazardous Waste Facilities through Alternative Systems for Mitigating and Managing Risk," *Hazardous Waste No. 3* (Fall 1984).

51. Michael Greenberg and Richard Anderson, *Hazardous Waste Sites: The Credibility Gap* (New Brunswick, N.J.: Center for Urban Policy, 1984).

52. Morell and Magorian, *Siting Hazardous Waste Facilities*.

53. Charles Davis, "Substance and Procedure in Hazardous Waste Facility Siting," *Journal of Environmental Systems*, No. 1 (1984–85).

54. The preceding case discussion is based on Daniel Mazmanian, Michael Stanley-Jones and Miriam Green, *Breaking Political Gridlock* (Claremont, Calif.: California Institute of Public Affairs, 1988).

55. Daniel Mazmanian and David Morell, "The Elusive Pursuit of Toxics Management," *The Public Interest*, No. 90 (Winter 1988).

56. Mazmanian, et al., *Breaking Political Gridlock*.

57. Lawrence Herson, "The Lost World of Municipal Government," *American Political Science Review* (June 1957).

58. Walter Rosenbaum, *Environmental Politics and Policy*, 2nd ed. (Washington, D.C.: CQ Press, 1991), p. 203.

59. Doyle, p. 38.

60. Elder Witt, "Getting Ready for the Big Spill," *Governing* (April 1988). p. 24.

61. Leslie R. Alm and Gary Silverman, *Hazardous Wastes and Materials Guidebook* (Bowling Green, Ohio: Center for Governmental Research and Public Affairs, 1991), pp. 16–17.

62. Ibid.

63. Denise Scheberle and Frank Lancaster, "Managing Small Quantities of Hazardous Wastes: The Emerging Role of Local Governments." Paper presented at the 1990 Annual Meeting of the American Society for Public Administration, Region VIII, Cheyenne, Wyoming.

64. George F. Gramling III and William L. Earl, "Cleaning Up after Federal and State Pollution Programs: Local Government Hazardous Waste Regulation," *Stetson Law Review, 17* (1988), pp. 639–87.

65. U.S. Senate, Committee on Environment and Public Works, *The Impact of the Proposed EPA Budget on State and Local Environmental Programs. Hearings before the Subcommittee on Toxic Substances and Environmental Oversight*, February 16 and March 28, 1983.

66. James P. Lester, "New Federalism and Environmental Policy," *Publius, 16* (Winter 1986).

67. U.S. Environmental Protection Agency, *Activities of EPA Assistance Programs* (Washington, D.C.: Government Printing Office, 1990).

68. Association of State and Territorial Solid Waste Management Officials, *State Programs for Hazardous Waste Site Assessments and Remedial Actions*. (Washington, D.C.: ASTSWMO, 1987).

69. Data obtained from a survey conducted by Professor Richard Feiock, Florida State University, Tallahassee, Florida.

70. EPA, *An Analysis of State Superfund Programs*, p.13.

71. Ibid., pp. 20–24.

72. Ibid., p. 24.

73. James K. Hammitt and Peter Reuter, "Illegal Hazardous Waste Disposal and Enforcement in the United States," *Journal of Hazardous Materials* (September 1989).

74. U.S. Environmental Protection Agency, *FY 1989 Enforcement Accomplishments Report* (Washington, D.C.:, Government Printing Office, 1989).

75. Ibid.

76. Goetze and Rowland, "Explaining Hazardous Waste Regulation at the State Level."

77. Lester, et al., "Hazardous Wastes, Politics and Public Policy."

78. State enforcement data were obtained from EPA under the authority of the Freedom of Information Act. Other state characteristics were obtained from standard sources such as government documents.

79. Charles Davis, Susan Kirkpatrick, and Denise Scheberle, "Evaluating Hazardous Waste Policymaking in Western States." In Zachary Smith, ed., *Environmental Politics and Policy in the West* (College Station: Texas A&M University Press, 1992).

80. Lester, et al., "Hazardous Wastes, Politics and Public Policy."

81. Seymour Schwartz, Wendy Cuckovich, Cecilia Fox, and Nancy Ostrom, "Improving Compliance with Hazardous Waste Regulations among Small Businesses," *Hazardous Wastes and Hazardous Materials*, 6 (Summer 1989).

82. Ibid.

6

HAZARDOUS WASTE POLICY FEEDBACK AND EVALUATION

Hazardous waste policies have undergone considerable change since their enactment by Congress over a decade ago. Much more is known now about the scope of pollution problems and about ways of minimizing or controlling wastes. Political institutions at all levels of government have grappled with the necessity of applying new policies to an expanding universe of regulated parties without a corresponding increase in staff or financial resources. Each jurisdiction is confronted with level-specific challenges.

EPA is ultimately responsible for the attainment of policy goals. Hence, agency officials must provide a usable policy and regulatory framework for the resolution of waste management problems. State policymakers are saddled with the politically contentious task of finding sufficient waste-disposal capacity to meet the needs of home state industries, while local officials must establish land-use policies affecting facility siting decisions and the management of underground storage tanks. Industry officials are increasingly asked to consider taking a proactive stance toward waste management on top of existing requirements mandated under RCRA or Superfund.

While political institutions and regulated parties properly remain the center of attention in our evaluation of hazardous waste policymaking, it also is important to direct attention to political processes and resource needs. Both intergovernmental and intersectoral relationships affect levels of understanding and cooperation among organizational decision makers, which are critical to the management of program activities. To better ascertain who needs to do what at a given degree of effort requires more and better information about staffing patterns in relation to various tasks.

In this chapter, the performance of these institutions is evaluated in relation to both political and administrative criteria. This evaluation is partially based on the conclusions reached thus far. Additional materials containing evaluative or prescriptive information are integrated where appropriate. We conclude with a discussion of policy dilemmas that are likely to affect the subsequent reauthorization of RCRA and Superfund.

INSTITUTIONS

Federal

Members of Congress took differing policy routes to enact hazardous waste laws after the political momentum for tough environmental laws peaked in the early 1970s and had begun to decline as a priority-issue area. RCRA achieved agenda status in 1976 largely through the efforts of policy specialists within EPA and congressional staff rather than through initiatives taken by environmental interest groups or the president. The law itself was not prominently marketed as a major pollution-control statute but was instead crafted as a section within the Solid Waste Disposal Act. Controversy was low to nonexistent, and RCRA was clearly overshadowed as a political issue by the Toxic Substances Control Act, which served to energize environmental activists and representatives of the Chemical Manufacturers' Association.

Conversely, Superfund carried a higher, more visible political profile as a consequence of the well-publicized toxic spills at Love Canal, New York, and related incidents across the United States. Public opinion, media attention, the involvement of President Carter, internal squabbling within the ranks of industries with a financial stake in the proposed policy, and intense lobbying by environmental groups contributed to the adoption of a new pollution-control law despite poor economic conditions and growing concern about regulatory compliance costs.

In the early 1980s, new information pertaining to the increasing magnitude of waste-disposal problems surfaced, along with the accompanying realization that existing laws were either flawed or silent in addressing important policy issues. Congress responded to these concerns in the reauthorization of both RCRA and Superfund by eliminating loopholes, strengthening EPA's enforcement authority, and increasing budgetary resources. Its distrust of President Reagan's political appointees at EPA and elsewhere was reflected in the inclusion of important statutory changes that reduced decision-making autonomy within EPA. Greater emphasis was placed on the use of goals and timetables as an action-forcing device to speed regulatory development and the incorporation of public participation opportunities as a means of reinforcing existing mechanisms of agency oversight.

Federal legislators also sought to communicate the need for a more aggressive regulatory stance to EPA through oversight hearings. Several of these were held during the final year of Anne Burford's tenure as EPA administrator. Information was disclosed leading to the dismissal of high-level officials and a number of statutory deficiencies were highlighted in the process. In 1985, hearings were held on the enforcement of groundwater monitoring requirements, a key RCRA provision. An important result was the issuance of a strongly worded recommendation that EPA pay closer attention to state administrative actions.[1]

Another area of responsibility involved matching resources to program needs. Federal hazardous waste programs took a sizable budget cut in the early 1980s, reflecting President Reagan's preference for lessening regulatory compliance costs within the private sector. Certain line items such as enforcement or research and development were especially hard-hit. Spending levels gradually rose from the mid-1980s to the present because of greater support for the attainment of policy goals expressed by EPA leadership and legislators from both sides of the aisle. Despite concern about the fiscal implications of expanding environmental programs voiced by President Reagan and OMB Director James Miller, Congress enacted HSWA and Superfund with a significant boost in monetary resources. Yet another increase was achieved at the request of EPA Administrator William Reilly, who sought to strengthen the enforcement of Superfund.

Some attempt has thus been made to compensate for the leaner hazardous waste program budgets of the early 1980s. This has been achieved in the face of constraints on agency spending imposed by the Gramm-Rudman-Hollings Act of 1985. Nevertheless, congressional efforts have been criticized for giving short shrift to HSWA if resource needs are compared to the rise in administrative responsibilities mandated under the law and consideration is given to the relatively short time frame allowed for the development of regulations and the issuance of permits.

EPA analysts also contend that resource constraints are jeopardizing program performance in less direct but equally significant ways, such as an overdependence on contractors for ongoing management activities. This has produced potentially negative consequences for quality control and accountability and an inability to retain skilled administrators within the RCRA program because of low salaries and a lack of career advancement opportunities.[2]

Congress thus stepped in to fill a policymaking void, since the Reagan Administration displayed little interest in offering new legislative proposals. Institutional strengths include policy development and oversight plus an ability to plug statutory loopholes in response to new information. It also appears that congressional actions writ large have been more responsive to public opinion on hazardous waste issues than to presidential, partisan, group, or regional economic interests.

Critical comments on legislative performance generally have centered on the gap between EPA's growing administrative workload and the provision of the necessary resources to do the job.[3] Another shortcoming on the policy side is the failure to address the problem of waste disposal within the federal government—primarily the Departments of Defense (DOD) and Energy (DOE)—by giving EPA greater authority to police waste-disposal problems found at other agencies.[4]

While Congress chose to keep EPA on a relatively short leash, agency administrators confront other political and institutional actors that affect and in some cases diminish decision-making autonomy. Presidential policy priorities clearly shape organizational choices and resources through the appointment of the EPA administrator and key positions within the White House staff, the OMB, and other executive departments. During the Carter presidency, EPA fared well in terms of executive and fiscal support for hazardous waste policymaking. Carter was personally involved in many of the strategic political decisions that eventually contributed to the enactment of Superfund.

EPA as an agency enjoyed precious little policy continuity after the inauguration of President Reagan. Pollution-control programs, including RCRA and Superfund, were increasingly evaluated less in terms of protecting environmental quality and public health than as a barrier to the attainment of economic growth. Budgets were slashed, appointees displayed greater loyalty to overarching administration policy goals than to those more central to EPA's mission, and administrative procedures were changed to tighten OMB control over agency decisions. Despite mounting evidence that other federal agencies such as the DOD and DOE flagrantly violated RCRA requirements, EPA was not given the necessary authority to compel compliance with the law. Through a pair of executive orders issued by Reagan, OMB gained the authority to review EPA's annual regulatory calendar and impact analyses attached to proposed regulations.

A consequence of these changes—in a cumulative sense—is a rather sizable impact on policy resulting from the transition from one presidency to the next. Perhaps the most important factor that ultimately shapes the implementation of hazardous waste programs is the appointment of the EPA administrator. The administrator has been well positioned to do this because of opportunities to establish policy or administrative priorities and communicate them to field administrators, state officials, and regulated parties; develop links with important agency constituencies; lobby key people within congress and the administration for the laws and resources needed to achieve program goals; and maintain a visible and credible regulatory presence through a combination of enforcement actions and media access. Various EPA administrators have excelled at one or more of these skills in working toward the attainment of policy goals.[5]

Other factors can be identified as contributors or barriers to the implementation of hazardous waste programs. Public opinion has consistently worked in favor of those advocating stricter legislation and offers a more suitable political climate for statutory enforcement as well. Congressional oversight hearings have led to closer EPA scrutiny of state efforts to administer RCRA. And policy design characteristics such as the goals and timetables approach incorporated within HSWA and SARA have led to observable increases in both enforcement actions and NPL site cleanup decisions. Obstacles to the realization of policy goals include institutional constraints such as OMB scrutiny of EPA's regulatory proposals, the technical complexity of project decision making on Superfund sites, and the occasional failure of high-level officials to offer sufficient amounts of policy support or resources.

How, then, can EPA's performance be evaluated? Within the realm of policy development, agency contributions have varied considerably according to the role assigned by different Presidents. EPA Administrator Russell Train was actively involved in putting together the RCRA package in the mid-1970s, while his successor, Douglas Costle, was instrumental in selling Superfund to members of Congress as a public-health issue. Reagan appointees Anne Burford, William Ruckelshaus and Lee Thomas were not encouraged to take a proactive policymaking stance.

While no new policy initiatives have been enacted during President Bush's term in office, the current EPA administrator, William Reilly, has demonstrated a willingness to adopt a more active role in promoting agency priorities. Whether he eventually succeeds will depend, in part, on evaluations of his efforts to overcome opposition to regulatory initiatives from the White House staff, the newly established Competitiveness Council (headed by Vice President Dan Quayle) and the OMB.

Four other important decisional arenas include the coordination of federal-state relationships, regulatory development, the communication of policy objectives, and enforcement. Again, EPA performance has varied across and within administrations. Under President Carter, the process of granting partial or full primacy to the states was quite slow at first, owing in large part to the need to have regulations in place. Since the first wave of basic rules to implement RCRA did not become effective until 1980, the final year of the Carter presidency, it became inevitable that incoming EPA administrator Anne Burford would be well positioned to process the backlog of state applications. As a strong supporter of program decentralization, she was eager to oblige—by the end of 1983, most states had received authority to administer RCRA. State officials were generally pleased by the opportunity to assume more responsibility, but their enthusiasm was tempered by administration efforts to reduce or eliminate grant monies designed to supplement operating costs.

In 1984, EPA again applied the brakes to the delegation of program management decisions. The newly adopted Hazardous and Solid Waste Amendments called for strict new changes in the authorization process—notably a requirement that the delegation of decision-making authority be withheld until regulations equivalent to those promulgated by the federal government were formally approved by state agencies. Equivalency was essentially checked by federal officials, who went through a line-by-line comparison of the two sets of rules. EPA regulations were to remain in effect until state regulatory actions were completed. Since then, relatively few states have been authorized to administer the RCRA program.[6]

Whether one views the current trend as beneficial or problematic is partly a philosophical question revolving around preferences for state versus federal control. The intent of the more rigorous clearance process is to ensure that a meaningful regulatory floor is established at the state level. This, in turn, makes it more difficult for industry officials to wave the relocation threat in front of state regulatory officials in retaliation for the expected increase in compliance costs. Another goal is making certain that states are in fact prepared to assume program management responsibilities. Thus, EPA is attempting to minimize the possibility of authorizing "paper programs."

A number of critics, including many state officials, contend that the effect of these changes is to subvert the very notion of a federal-state "partnership."[7] A statutory approach such as primacy masks the reality of federal control with the appearance of flexible intergovernmental arrangements that are allegedly sensitive to differing conditions within each state. More specific complaints include the failure of federal officials to clarify how states are to resolve potential conflicts between authorizing criteria—consistency, stringency, and equivalence.[8] Would strict adherence to these criteria leave any room for state officials to try something different?

Questions also have been raised about the compatibility of the existing process of delegating program authority with the HSWA statute. Since HSWA prescribes the use of goals and timetables for its numerous regulatory provisions, some argue that it is unrealistic to expect state regulatory officials to respond in near lockstep form to EPA administrators who are themselves struggling to keep up. The task of playing regulatory catch-up at the state level is doubly challenging if the political need to obtain both resources and legal authority from the state legislature is considered.

In short, EPA has been criticized for its failure to clarify what state officials must do to construct a passable application form for the management of RCRA. However, Congress also has contributed to this problem by adopting legislation that combines a decided lack of discretionary authority for EPA with a short time frame for regulatory development.

EPA's performance within the regulatory arena can be assessed on two counts: the development of rules in a timely fashion and the issuance of permits. Rulemaking has not proven to be one of the agency's strong suits. Despite a congressional deadline of 1978 for the promulgation of basic regulations for RCRA, EPA did not put forward the initial set until 1980. These efforts were noticeably slowed by an array of factors, including a lack of precedent, statutory ambiguity, technological uncertainty, resource constraints, and within-agency disputes over the shape and direction of proposed rules. Unfortunately, the timeliness of regulatory production has not improved since the enactment of HSWA in 1984. Tardiness in this case has been attributed to two interrelated factors: insufficient staffing levels in relation to mandated responsibilities and high employee turnover.[9]

EPA fares better in the category of issuing permits. Agency administrators were faced with an altered set of management tasks after Congress enacted HSWA in 1984. On the one hand, the regulatory universe increased considerably. Nearly 5,000 new facilities requiring an operating permit for the processing, storage, or disposal of hazardous wastes were added. This seemingly daunting challenge was negated by a countervailing shift—the strengthening of environmental quality standards. A short-term consequence of these changes was the closure of approximately two-thirds of these facilities throughout the United States because of the inability of owners to cover the additional compliance costs.

This left EPA with the responsibility for making permit decisions on the remaining firms within the deadlines established by Congress.

This task has been fulfilled for the most part. Over 90 percent of land disposal and incinerator facilities were issued permits on time. Moreover, this occurred within a decision-making process that resulted in the denial of a significant percentage of applicants (10 percent), thus suggesting that efficiency was not pursued without regard for quality control objectives.[10]

A third area of concern is the communication of policy objectives. This includes contacts between the regulators and members of the regulated community, EPA headquarters, and the regional offices and state and federal program administrators. In the early 1980s, EPA Administrator Burford made little headway in promoting good relationships with either the private sector or state agencies. Initial gestures consisted of a pledge to begin and maintain a nonadversarial relationship with industry. This resulted in the transmission of an unintentional message to waste-generating firms or tsd facilities that the risks associated with ignoring hazardous waste regulations were relatively slight.

When Burford was succeeded by William Ruckelshaus as EPA administrator, an effort was made to shift gears. He made selective use of well-publicized lawsuits against egregious offenders of hazardous waste policies as a means of deterring noncompliance. Since then, deterrence rather than promises of harmonious working relationships has represented the preferred means of inducing regulated parties to comply with RCRA and Superfund.

Communications problems are found within the realm of regulatory interpretation as well as enforcement. Both industry and state officials want EPA to clarify ambiguous statutory or regulatory language. Examples include an easily understandable interpretation of EPA's definition of hazardous waste, the practical distinction between "guidelines" and "directives," and the amount of weight assigned to potentially conflicting decision-making criteria such as equivalence and stringency.

Regulatory uncertainty has led industry officials to undertake a certain amount of "forum shopping" across differing levels of government to identify legally defensible courses of action. Poor communications between federal and state administrators also lead to duplication of tasks, accusations of second-guessing, and greater delay in making or reviewing decisions.[11]

Another source of tension affecting the intergovernmental management of hazardous waste programs is derived from differing views of who ought to be responsible for what. State administrators take a page from much of the "New Federalism" rhetoric, which identifies the states and the federal government as partners in the resolution of policy problems, while actual working arrangements can more easily be depicted as a contractor-client relationship. Even though EPA administrators continue to experiment with new approaches to the implementation of RCRA and Superfund, state officials are rarely consulted as fellow professionals with something to offer on matters affecting policy development or enforcement. There is preliminary evidence that indicates that efforts to build in more frequent contacts between state and regional EPA officials are rewarded with more efficiently run programs.[12]

A fourth area of concern for EPA is enforcement. The approach taken has evolved from a less adversarial, more cooperative relationship with industry to a focus on deterrence, including stiffer criminal and civil penalties for noncompliance. With Superfund, this has meant devoting more attention and resources to the permanent cleanup of contaminated NPL sites relative to emergency response actions. More effort also has been directed toward the identification of potentially responsible parties (PRPs) to contribute their share of cleanup costs, thereby allowing EPA to stretch its resources further. This is the agency response to one of the chief complaints offered by critics; i.e., getting more sites decontaminated and de-listed.

Accelerated action in the cleanup of NPL sites was prompted by the establishment of performance deadlines by Congress within the statutory language of SARA. Whether these goals can be achieved in a timely fashion is debatable. An important technical constraint responsible for much of the delay in the decontamination of sites is the procedural complexity of decision making. A related administrative problem is EPA's reluctance to establish policy guidelines for the identification of top-priority sites for initial cleanup work.[13]

RCRA enforcement is more of a joint enterprise between federal and state administrators. EPA's responsibilities include the inspection of waste handlers, oversight of state and local regulatory efforts and, if necessary, initiation of enforcement proceedings against noncompliant firms. Agency performance here has evolved in much the same fashion as with the Superfund program. Buttressed by the support of agency directors more favorably predisposed toward a strong enforcement posture, an increase in legal enforcement authority granted by Congress under HSWA, and more resources, EPA has achieved gradual progress in carrying out enforcement-related activities following the regulatory lull of the early 1980s.

EPA also has assumed important new regulatory tasks related to the problem of noncompliance with RCRA occurring at waste management facilities. Agency administrators oversee corrective actions undertaken by facility operators to address these problems, including the instigation of preventive measures designed to reduce the likelihood of future waste leaks and the cleanup of wastes that have already migrated through soil or, in some cases, groundwater.

While EPA has quickened the pace of RCRA implementation in terms of aggregate enforcement statistics, the numbers per se do not necessarily indicate whether a deterrent effect is occurring. One shortcoming attributed to agency actions taken in the mid-1980s is that high-priority violators were less likely to have been dealt with in an appropriate and timely manner than low-priority violators. Another critique focused on EPA's neglect of the oversight function, particularly the politically unpleasant task of intervention to assert regulatory authority because of state inaction.[14]

To summarize, federal institutions such as Congress and EPA have done reasonably well in forging policy or administrative responses to waste management problems and in adapting to changing circumstances through the incorporation of new information into the reauthorization of RCRA and Superfund. More critical evaluations of institutional performance typically center on the mismanagement of specific program responsibilities, the identification of statutory loopholes not yet addressed, and a failure to consider the resource requirements associated with policy mandates. A more complete assessment of legislative or executive accomplishment awaits the development of better indicators of program success.

State and Local

State and local environmental agencies had no experience in the management of hazardous waste programs prior to the federal enactment of RCRA and Superfund. Since then, state policymakers have made significant progress in passing companion legislation allowing state agencies to regulate firms involved in the generation, processing, transport, storage, or disposal of wastes. Most states have received authorization from EPA to manage the basic regulatory package associated with RCRA, but relatively few have achieved primacy for the HSWA program. In addition, nearly all states have adopted Superfund programs to aid in the decontamination of sites that are laden with pollutants but are too small to qualify for the federal NPL. While a few states have been quite innovative in creating new policy approaches for the management, recovery or disposal of hazardous wastes, the majority have been content to produce laws with the minimum content needed to obtain program management authority.

Experimentation is most likely to occur in policy areas clearly reserved for subnational governments, namely the siting of hazardous waste facilities. Several different policy approaches have been tried, ranging from those of strong state control over the decision-making process to those that treat the presence or absence of tsd facilities as nothing more than another community land-use decision already covered in city or county laws. Thus far, there is little evidence to suggest that any design is more likely to result in a sited facility than others.[15] Political factors such as NIMBY continue to pose a formidable barrier to site development, while a corresponding interest in policy proposals that treat facility siting as a problem requiring consensus-building efforts in communities across the state has surfaced in some jurisdictions.[16]

Local governments also have become more active in the adoption of hazardous waste legislation, partly in response to the federal Community Right to Know Act (Title III of SARA) and partly because of increasing awareness of waste-disposal problems posing unacceptable risk to public health or environmental quality. Facility siting controversies have certainly energized residents in numerous communities across the United States, prompting several larger chemical firms to pay closer attention to source reduction as an alternative to off-site disposal of toxic wastes.

Another issue of concern to local officials, particularly fire department administrators, is the need to extract or neutralize the risk of contamination originating from leaking underground storage tanks containing petroleum or other hazardous chemicals. Some cities have recognized the importance of protecting groundwater as a major source of municipal supply and have enacted laws that affect waste-disposal options; others have restricted facility development or on-site storage through zoning restrictions; and a few have adopted ordinances requiring proof that property is free of pollutants before it can be sold.[17]

The availability of fiscal and staff resources has plagued state and local governments since the inception of hazardous waste policy. Early on, state agencies began receiving grants from EPA for program operations. However, it became evident that the degree of commitment varied widely across the states. Several larger jurisdictions saddled with serious problems of groundwater and land contamination anted up the necessary appropriations to supplement federal monies in developing an effective regulatory program, while others were largely dependent on EPA grants to mount any sort of response to pollution problems.

From the mid-1980s to the present, state spending for HSWA and Superfund activities has risen steadily. Much of this increase has come from general appropriations, but user fees have become a common means of raising revenue for program expenses as well.[18] Unfortunately, the discovery of new contaminated sites as well as corrective action expenditures for existing facility operations means that the resource-to-problem gap may be wider rather than narrower.

While many states have found it difficult to find adequate own-source funding for RCRA and Superfund, the resource problem is not likely to improve. The mounting cost of liability insurance (if it is available) has become quite troublesome for both industry and tsd facility owners nationwide, resulting in the consideration of policies that put the state in charge of managing financial risk. Another problem is a chronic shortage of trained personnel to handle waste management activities. Whether this challenge can be met by relying more extensively on private contractors or an influx of graduates from new university or community college programs in hazardous waste management remains to be seen.

The implementation of hazardous waste policies at the state and local level is somewhat uneven in a geographical sense. Our findings indicate that the likelihood of inspecting or citing a tsd facility for violations is considerably greater in southern or northeastern states than in other regions. There also are political and economic characteristics that differentiate states on the basis of regulatory effort. By far, the most important factor that accounts for the initiation of enforcement actions is problem magnitude. Regulatory officials are more likely to issue administrative orders in states that generate more waste and are home to more NPL sites.

Other factors contributing to stronger enforcement efforts include the relative comprehensiveness of state hazardous waste statutes, stronger voter identification with the Democratic party, and an economic dependency on the production of chemicals as a source of state jobs. Enforcement of hazardous waste laws, particularly HSWA, at the local level is more troublesome because of the inability or unwillingness of policymakers to address the problem of widespread noncompliance by small firms operating at the margins of economic profitability.

What does this tell us about the ability of state and local governments to wrestle with hazardous waste problems? Generally speaking, they have made progress in developing the necessary policy infrastructure to deal with waste-disposal problems. States have enacted legislation that faithfully replicates the intent of Congress in calling for the regulation of waste-generating firms and the cleanup of contaminated sites.

However, policymakers have not been successful either in resolving facility siting issues or in preventing the overly rapid depletion of in-state waste-disposal capacity through bans or restrictions on the importation of hazardous wastes generated elsewhere. Nor have public officials taken the initiative to augment federal monies targeted for regulatory activities with own-source revenues or to invest in the kinds of educational or training programs needed to alleviate personnel shortages within these programs.

UNRESOLVED ISSUES

Resources

The interplay between resource and policy needs is likely to dominate forthcoming debates in Congress over the reauthorization of RCRA and Superfund. Studies by EPA, university researchers, OTA, GAO, and CBO are unanimous in concluding that HSWA and SARA are underfunded in relation to what agencies need to do to achieve legislative goals.[19] While Congress and state legislatures have gradually increased expenditures for program operations (typically within an austere fiscal climate), EPA and state environmental agencies are still hard-pressed to meet statutory deadlines for regulations and to carry out enforcement and corrective action decisions. Administrators argue that more resources are required not only for basic operating expenses but for incentive packages to hire and retain qualified staff and for improvements in the quality of information needed to make hard choices in the selection of policy or management options.

The latter concern represents the kind of investment that could greatly add to our understanding of hazardous waste problems, according to policy evaluators who either work for EPA or study it. Both OTA and GAO have pointed out the wide variability of estimates in the national production of hazardous waste and the absence of data dealing with treatment, storage, or waste-disposal capacity.[20] Knowing in more precise terms the amount and location of waste generated and the state or regional availability of viable control options can provide political ammunition for state policymakers attempting to fashion a legislative solution for a lack of capacity assurance. Or it may offer a stronger data-based rationale for regional EPA officials who must decide whether withholding Superfund monies is a useful means of prodding waste-exporting states to assume responsibility for more of the pollution-control costs generated within their borders.

Better information also could hasten the development of better indicators of policy impact, such as the extent to which hazardous waste programs have contributed to improvements in environmental quality or a reduction in health risks for people residing near waste-generating firms or tsd facilities. This in turn would enable Congress, the president, environmental groups, and others to assess agency performance as well, thereby enhancing the prospects for organizational accountability.

Finally, EPA or state departmental administrators might seize the opportunity to identify both training and personnel requirements needed to achieve assigned responsibilities. GAO has suggested that Congress place the statutory responsibility for gathering and processing information in EPA's lap and provide the necessary legal and financial resources to overcome the rather sizable between-state differences in data-acquisition efforts.[21]

Liability

One of the more contentious sets of unresolved issues is contained under the topic of liability. This term implies several things within the context of hazardous waste policymaking. In its most direct form, liability refers to the failure of organizations to meet their legal obligations. Federal departments such as the DOD and DOE have been criticized for not making a more serious effort to comply with RCRA and Superfund. Some have suggested that the appropriate congressional response to environmental neglect of this sort is to give EPA or the Justice Department more direct regulatory authority over other executive agencies.[22] Another approach that has received some attention in the corridors of Congress is the provision of line-item funding for departmental compliance with toxic-waste laws. The intent here is to ensure that an agency could meet its pollution-control obligations without sacrificing resources needed for the attainment of its own major policy goals.

Liability also is invoked in policy discussions dealing with the reauthorization of Superfund. The issue of who pays for the cleanup of abandoned hazardous waste dump sites has become particularly controversial. Representatives of banks and other financial institutions are dismayed by federal court decisions that hold them liable for the decontamination of properties acquired through foreclosure.[23] This interpretation of CERCLA, they argue, is a product of judicial creativity rather than congressional intent. An unfortunate consequence is increasingly restrictive lending practices, resulting in the construction of roadblocks for small businesses or individuals seeking loans. Industry officials typically prefer an alternative approach, such as a no-fault system based on the notion that program costs ought to be shared more widely.[24]

On the other hand, environmental leaders resist such initiatives because of the belief that incremental unraveling of the financial underpinnings of Superfund would open up the policy floodgates to other aggrieved parties (such as municipalities). Some are suspicious that proposals of this sort are motivated by a desire to dilute support for the law by placing less emphasis on cleanup efficiency questions than the fairness issue. Just as members of the environmental community won political skirmishes in the past by focusing on potent symbols such as public-health risks or midnight dumping, lenders have sought public support by suggesting that the survival of small banks is jeopardized by the actions of a large and uncaring bureaucracy. Aside from the merits of the individual claimants, environmental representatives are attempting to prevent industry tampering with the overarching theme etched into the original statute: the "polluter-pays" principle.[25]

Yet another liability-related concern is the availability and cost of insurance coverage for those who generate or handle hazardous wastes. Congressional architects of hazardous waste policy originally viewed the incorporation of pollution insurance requirements within HSWA and SARA as a potentially useful means of prodding tsd facility operators to pay more attention to safety and administrative concerns. An added bonus was the oversight of facility operations by a nongovernmental actor poised to take action if good management practices were not followed.[26]

However, few policymakers could have anticipated the dramatic rise in litigation that followed and the subsequent elimination of any economic incentive for insurers to offer coverage for hazardous waste spills. The number of firms writing insurance policies for hazardous waste has declined sharply since the early 1980s, and the expense of acquiring such insurance from the few remaining companies has skyrocketed. A GAO report estimated that the average cost of premiums rose elevenfold between 1982 and 1988.[27]

According to industry officials interviewed in a recent survey of major insurers, there are three major reasons cited for leaving this area of business. They include declining profitability precipitated by the spiraling costs of Superfund cleanups (around $29 million per NPL site), federal court decisions upholding a strict liability interpretation of hazardous waste statutes, and an increase in lawsuits initiated by client firms (such as banks) seeking to avoid or reduce financial loss.[28]

The lack of insurance has proven to be particularly troublesome for both federal and state officials, who have fielded numerous requests from regulated firms to come up with a policy solution. GAO has recommended that Congress authorize EPA to collect information detailing the amounts of indemnity payments made by insurance companies to cover NPL cleanup costs. Agency administrators could then determine whether sufficient financial capacity exists nationwide to meet statutory requirements.[29]

States have scrambled to address this problem as well. Many jurisdictions have responded by allowing financial responsibility tests in lieu of insurance, such as the use of corporate guarantees or a demonstration by facility owners that sufficient monetary assets are available to cover expenses incurred from an accidental release of toxic wastes.[30] This approach allows financially solvent medium to large firms a means of satisfying the legal requirements mandated under RCRA, but it does little to address the needs of small independent businesses such as gas station owners who lack the financial wherewithal to "pass" the fiscal responsibility test.

Waste Disposal Capacity

Acquiring the necessary capacity to manage or dispose of hazardous wastes generated within state borders is one of the most politically charged issues to be found on state or local governmental agendas. Put simply, it is a policy question that considers the redistribution of risk and cost. People living close to a newly constructed tsd facility must bear the economic and psychological burdens associated with a perceived decline in the quality of life, while proximate cities or towns with waste-producing firms receive the benefits of less costly waste-disposal options plus a lesser risk of illegal dumping.

The NIMBY syndrome continues to pose a significant political hurdle for policymakers seeking additional facility space. In addition, the pressure to find a policy solution has been exacerbated by a provision within SARA that authorizes EPA to withhold the release of Superfund monies to states that are net exporters of waste; that is, the amount of waste sent to facilities in other states exceeds the volume of toxic materials retained for treatment, storage, or disposal.

The politics of waste disposal contains both internal and external forms of expression. Lawmakers looking inward for a state solution to the policy impasse must inevitably come to grips with the need to site new facilities, although a more politically appealing approach is to call for the creation of waste-minimization strategies (discussed in the following section). While NIMBYism is difficult to overcome under any circumstances, there are both political factors and institutional forms that contribute to a less adversarial forum for site-selection decisions and the fashioning of a policy consensus.

The most important political factors identified by Morell and Magorian in their analysis of substate controversies in New Jersey included a belief among local residents that housing a facility would contribute to the economic health of the community without creating undue risk to public health.[31] Potentially useful institutional approaches include the development of a state or substate regional authority that is not perceived to be controlled by any particular group and a state selection process that is built on information contained in planning documents submitted by substate governments.[32] Whether either approach can work in the absence of a political crisis remains to be seen.

The outward form of political discourse can be observed in the interstate battles over waste-export issues. Large producer states, such as Massachusetts, North Carolina, and Pennsylvania often lack the necessary storage or disposal capacity to take care of their own wastes. Hence, much of the toxic garbage is sent to tsd facilities in other jurisdictions. Policymakers representing net importer states have become increasingly frustrated with their status as waste havens and have opted for legislative solutions to redress inequities in burden sharing.

Public officials in Alabama initially adopted a statute calling for restrictions on the importation of hazardous wastes from states lacking treatment or disposal facilities. This law was declared unconstitutional by the Alabama Supreme Court in July 1990. State lawmakers then responded by imposing a $72-per-ton fee on out-of-state wastes.[33] Similar laws focusing on taxes or fees as the preferred means of discouraging interstate waste shipments also were enacted in Louisiana and South Carolina. Each law has been challenged in the federal courts by industry officials seeking to preserve exports as a viable disposal option.

While these statutes have been attacked as an unconstitutional infringement on interstate commerce, the attorneys general of the recipient states continue to argue that such arguments are outweighed by public health and safety considerations.[34] Ironically, between-state conflict is largely fueled by the reluctance of EPA administrators to enforce SARA provisions requiring greater effort on the part of states to produce capacity-assurance plans (which would result in more within-state conflict). The dilemma facing federal officials thus represents a catch-22 on the surface but in fact means that the relevant decisions will be made by judges rather than administrators.

Waste Minimization

Perhaps the one area of hazardous waste management receiving something akin to universal acclaim is waste minimization, or the notion that reducing the volume and toxicity of waste is preferable to reliance on regulatory strategies for handling toxic byproducts. The idea is not particularly new. In 1976, EPA outlined a hierarchy for waste management options that listed, in order of preference, waste reduction, recycling, treatment, and land disposal. Agency administrators hoped that the cradle-to-grave regulatory philosophy established under RCRA would provide an incentive for regulated parties to more seriously consider waste-minimization strategies. However, the effect of hazardous waste regulations was just the opposite. Industries invested in pollution control rather than prevention.[35]

Congress took action in 1984 with the enactment of HSWA to place restrictions on the use of land disposal as a management option. EPA was asked to prepare and submit a report by October 1986 on the feasibility of increasing the status of waste-minimization activities within agency operations. The report placed considerable emphasis on the economic and environmental rationale for a strategy of this sort but stopped short of recommending a full-blown regulatory program.

It also was suggested that more baseline information be collected on the amount and toxicity of waste generated within the United States, the availability of treatment and disposal facilities, and the effects of existing waste-disposal practices on public health and environmental quality.[36] The case for emphasizing minimization would clearly receive a political boost if subsequent data analyses revealed a widening gap between waste production and disposal capacity.

Perhaps the most important recommendation contained within the report called for EPA to initiate a nonregulatory program to educate and ultimately convince industry officials that integration of waste-minimization practices into company operations would contribute to a cleaner environment and lower costs of production.[37] There was considerable agreement that a more prescriptive approach not only would be difficult to implement because of technical and economic obstacles but also would alienate some industry leaders with a sizable prior investment in pollution-control equipment. Moreover, EPA officials were not inclined to send a message to regulated parties that hopping onto the waste-minimization bandwagon would in any way absolve them of the responsibility to meet ongoing RCRA requirements.

EPA actions taken thus far have been consistent with the general goals spelled out in its 1986 report to Congress. Staff within the Office of Solid Waste have been assigned to work on waste minimization projects, and much of the preliminary work falls within the realm of information collection and dissemination. This includes the development of a training program through grants-in-aid to the states; the provision of technical assistance; identifying departmentwide opportunities for pollution-prevention practices in the federal government; and serving as a clearinghouse for research and technology transfer between industry, universities, states, and EPA.[38] A small number of states and pollution-generating firms also have initiated programs, but it is unlikely that large-scale investment will occur without a major push from the federal government.[39]

Greater impetus for the concept of waste minimization within waste-generating industries was created by the enactment of the Pollution Prevention Act of 1990. Congress authorized the expenditure of $16 million per year from fiscal years 1991 through 1993 for EPA to develop a program to promote source-reduction strategies. Part of the money was earmarked for matching grants to the states, thereby encouraging subnational policy initiatives.[40]

Other components of this law included the provision of technical assistance to affected companies, the creation of a training program and assorted guidance documents, and improved public access requirements for individuals or groups seeking data pertaining to the extent of source-reduction efforts undertaken on an industrywide basis. Affected industries also were required to report amounts of toxic chemicals released into the environment to EPA along with information about the types of process changes made and recycling efforts. Such data could then be evaluated to assess program effectiveness later on.[41]

HAZARDOUS WASTE AND THE POLICYMAKING PROCESS: IS IT DIFFERENT?

Each substantive area of public policy is unique in highlighting certain features or attributes of the policy process, just as its enactment and implementation serves to corroborate much of what we know about the policymaking cycle. Like other regulatory programs under the environmental protection umbrella, RCRA and Superfund rank high on technical complexity and the breadth of impact. Industries producing a wide array of products with vastly different modes of production are affected by hazardous waste regulations administratively, through restrictions on waste-disposal options, and economically, through increasing compliance costs. The incentive for both environmental and regulated interests to mobilize in an effort to influence policy or regulatory change is quite high. To a greater degree than is true for other policy areas, the key political battles affecting pollution-control programs are fought on administrative turf rather than in the hallways of Congress.

What makes hazardous waste different in a policymaking sense? More than other policy areas, this issue has been affected by public opinion. Exhibit A is the initial Superfund law, which benefited politically from a de facto partnership between dramatic and continuous media attention and subsequent polls reflecting strong public support. An even better example of the opinion-policy nexus is shown by the enactment of HSWA (in 1984) and SARA (in 1986) by strong bipartisan majorities in Congress, despite the opposition of a popular incumbent president who viewed both policies as overly expensive and intrusive.

Locally, public opinion works in a more direct fashion to form a potent political roadblock to the adoption of LULUs (locally unwanted land uses). Very few tsd facilities have been sited over the past decade. Thus, hazardous waste policy choices within EPA are shaped to some extent by a political pincer effect. Agency administrators develop broad regulatory guidelines to implement policies that enjoy widespread support. However, their efforts to carry out preferred management options such as insistence on adequate waste-disposal capacity at the state level are often thwarted by public opposition. Whether policymakers are up to the challenge of closing the gap between concentrated costs and dispersed benefits with some form of politically acceptable burden sharing remains to be seen.

NOTES

1. The importance of these hearings as a prompt for stronger federal oversight of state RCRA programs is emphasized in a recent EPA report. Consult U.S. Environmental Protection Agency, *The Nation's Hazardous Waste Program at a Crossroads: The RCRA Implementation Study* (Washington, D.C.: Government Printing Office, July 1990).

2. Ibid., Chapter 8.

3. Walter A. Rosenbaum, *Environmental Politics and Policy* (Washington, D.C.: CQ Press, 1991), especially pp. 229–31.

4. These problems are fully documented in a recent study prepared for the U.S. Senate Government Affairs Committee. See Congressional Budget Office, *Federal Liabilities Under Hazardous Waste Laws* (Washington, D.C.: Government Printing Office, May 1990).

5. Useful sources for earlier EPA appointees and their priorities in office include the chapters by J. Clarence Davies III and Steven Cohen in Norman Vig and Michael Kraft, eds., *Environmental Policy in the 1980s* (Washington, D.C.: CQ Press, 1984). Vig and Kraft also discuss EPA Administrator William Reilly and other Bush Administration appointees in the introductory chapter of their *Environmental Policy in the 1990s* (Washington, D.C.: CQ Press, 1990).

6. EPA, *The Nation's Hazardous Waste Program at a Crossroads*, p. 21.

7. See Cheryl H. Wilf, "Administrative Preemption and State Authority: Revisiting the Middle Ground Thesis," APSA Conference Paper, San Francisco, 1990; and n.a., *State-Federal Relations in the RCRA Regulatory Program* (Keystone, Colo.: The Keystone Center, June 1989).

8. Ibid.

9. EPA, *The Nation's Hazardous Waste Program at a Crossroads*, Chapter 4.

10. Ibid., Chapter 5.

11. The problem of intergovernmental communications in hazardous waste programs is discussed in Malcolm Goggin, Ann O'M. Bowman, James P. Lester, and Lawrence J. O'Toole, Jr., *Implementation Theory and Practice* (Glenview, Ill.: Scott, Foresman/Little, Brown, 1990), especially Chapter 3.

12. EPA, *The Nation's Hazardous Waste Program at a Crossroads*, Appendix B.

13. Two reports take EPA to task for insufficient attention to the establishment of policy guidelines or quantitative indicators of progress in meeting the legislative objectives of Superfund. They are General Accounting Office, *Superfund: A More Vigorous and Better Managed Enforcement Program Is Needed* (GAO/RCED-90-22, December 1989); and Office of Technology Assessment, *Are We Cleaning Up?: 10 Superfund Case Studies* (Washington, D.C.: Government Printing Office, June 1988).

14. General Accounting Office, *Hazardous Waste: Many Enforcement Actions Do Not Meet EPA Standards* (GAO/RCED-88-140, June 1988).

15. Richard N.L. Andrews, "Hazardous Waste Facility Siting: State Approaches." In Charles Davis and James P. Lester, eds., *Dimensions of Hazardous Waste Politics and Policy* (Westport, Conn.: Greenwood Press, 1988); and Margaret Condron and Dixie Sipher, *Hazardous Waste Facility Siting: A National Survey* (Albany, N.Y.: Legislative Commission on Toxic Substances and Hazardous Wastes, June 1987).

16. Patrick G. Marshall, "Not in My Back Yard!" *Editorial Research Reports* (Washington, D.C.: Congressional Quarterly, Inc., June 9, 1989).

17. George F. Gramling III and William L. Earl, "Cleaning Up after Federal and State Pollution Programs: Local Government Hazardous Waste Regulation," *Stetson Law Review*, *17* (1988), pp. 639-87.

18. U.S. Environmental Protection Agency, *An Analysis of State Superfund Programs* (Washington, DC: Government Printing Office, 1989, pp. 6–7.

19. EPA, *The Nation's Hazardous Waste Management Program at a Crossroads*, Chapter 8; Rosenbaum, *Environmental Politics and Policy*; CBO, *Federal Liabilities*

Under Hazardous Waste Laws; Office of Technology Assessment, *From Pollution to Prevention: A Progress Report on Waste Reduction* (Washington, D.C.: Government Printing Office, June 1987); and General Accounting Office, *Hazardous Waste: EPA's Generation and Management Data Need Further Improvement* (GAO/PEMD-90-3, February 1990).

20. Office of Technology Assessment, *Serious Reduction of Hazardous Waste* (Washington, D.C.: Government Printing Office, September 1986; and General Accounting Office, *Hazardous Waste: New Approach Needed to Manage the Resource Conservation and Recovery Act* (GAO/RCED-88-115, July 1988).

21. GAO, *Hazardous Waste: EPA's Generation and Management Data Need Further Improvement.*

22. GAO, *Hazardous Waste: New Approach Needed to Manage RCRA*, pp. 46–47.

23. Dinah Wisenberg, "Threat of 'Superfund' Liability Makes Banks Wary of Loans," *CQ Weekly Report* (June 30, 1990), pp. 2046–47.

24. Peter W. Huber, *Liability: The Legal Revolution and Its Consequences* (New York: Basic Books, 1988).

25. Wisenberg, "Threat of 'Superfund' Liability," p. 2046.

26. General Accounting Office, *Hazardous Waste: Issues Surrounding Insurance Availability* (GAO/RCED-88-2, October 1987).

27. General Accounting Office, *Hazardous Waste: The Cost and Availability of Pollution Insurance* (GAO/PEMD-89-6, October 1988).

28. Ibid., Chapter 2.

29. Ibid. See also General Accounting Office, *Hazardous Waste: Pollution Claims Experience of Property / Casualty Insurers* (GAO/RCED-91-59, February 1991).

30. GAO, *Hazardous Waste: Issues Surrounding Insurance Availability*, p. 28.

31. David Morell and Christopher Magorian, *Siting Hazardous Waste Facilities: Local Opposition and the Myth of Preemption* (Cambridge, Mass.: Ballinger, 1982).

32. Marshall, "Not in My Backyard!" p. 316. See also Albert Matheny and Bruce Williams, "Strong Democracy and the Challenge of Siting Hazardous Waste Disposal Facilities in Florida," *National Civic Review* (July/August 1988).

33. Ronald Smothers, "States Battle U.S. and Other States Over Waste," *The New York Times* (January 27, 1991).

34. Sue Darcey, "States Fight to Ban Out-of-State Wastes," *World Wastes, 33* (March 1990), pp. 36–38.

35. GAO, *Hazardous Waste: New Approach Needed to Manage RCRA*, Chapter 4.

36. U.S. Environmental Protection Agency, *Minimization of Hazardous Waste* (Washington, D.C.: Government Printing Office, October 1986).

37. Ibid.

38. GAO, *Hazardous Waste: New Approach Needed to Manage RCRA*, pp. 54–56.

39. OTA, From *Pollution to Prevention*. See also Joel Hirschhorn, "Preventing Industry Waste," *EPA Journal* (January/February 1990), pp. 36–39.

40. Congressional Quarterly, Inc., *CQ Almanac 1990* (Washington, D.C.: Congressional Quarterly, Inc., 1991), p. 155.

41. Ibid.

REFERENCES

BOOKS

Brown, Michael. *Laying Waste: The Poisoning of America by Toxic Chemicals*. New York: Pantheon, 1980.

Brown, Phil, and Edwin J. Mikkelsen. *No Safe Place: Toxic Waste, Leukemia and Community Action*. Berkeley: University of California Press, 1990.

Cohen, Steven, and Sheldon Kamieniecki. *Environmental Regulation through Strategic Planning*. Boulder, Colo.: Westview Press, 1991.

Davis, Charles E., and James P. Lester, eds. *Dimensions of Hazardous Waste Politics and Policy*. Westport, Conn.: Greenwood Press, 1988.

Doniger, David D. *Law and Policy of Toxic Substances*. Baltimore, Md.: Johns Hopkins, 1979.

Epstein, Samuel S., Lester O. Brown, and Carl Pope. *Hazardous Waste in America*. San Francisco: Sierra Club, 1982.

Fortuna, Richard, and David Lennett. *Hazardous Waste Regulation: The New Era*. New York: McGraw-Hill, 1987.

Gibbs, Lois Marie. *Love Canal: My Story*. Albany: State University of New York Press, 1982.

Goldman, Benjamin. *Hazardous Waste Management: Reducing the Risk*. Washington, D.C.: Island Press, 1986.

Greenberg, Michael R., and Richard F. Anderson. *Hazardous Waste Sites: The*

Credibility Gap. New Brunswick, N.J.: Center for Urban Policy Research, 1984.

Harthill, Michalann, ed. *Hazardous Waste Management: In Whose Backyard?* Boulder, Colo.: Westview Press, 1984.

Hirschhorn, Joel, and Kirsten Oldenburg. *Prosperity Without Pollution*. New York: Van Nostrand Reinhold, 1991.

Landy, Marc, Marc Roberts, and Stephen Thomas. *The Environmental Protection Agency: Asking the Wrong Questions*. New York: Oxford University Press, 1990.

Lehman, John, ed. *Hazardous Waste Disposal*. New York: Plenum Press, 1983.

Lester, James P., and Ann O'M. Bowman, eds. *The Politics of Hazardous Waste Management*. Durham, N.C.: Duke University Press, 1983.

Levine, Adeline G. *Love Canal: Science, Politics, and People*. Lexington, Mass.: Lexington Books, 1982.

Mazmanian, Daniel, and David Morell. *Beyond Superfailure: America's Toxics Policy for the 1990s*. Boulder, Colo.: Westview Press, 1992.

Morell, David L., and Christopher Magorian. *Siting Hazardous Waste Facilities: Local Opposition and the Myth of Preemption*. Cambridge, Mass.: Ballinger, 1982.

Nader, Ralph, Ronald Brownstein, and John Richard, eds. *Who's Poisoning America: Corporate Polluters and Their Victims in the Chemical Age*. San Francisco: Sierra Club, 1981.

O'Hare, Michael, Lawrence Bacow, and Debra Sanderson. *Facility Siting*. New York: Van Nostrand, 1983.

Piasecki, Bruce, ed. *Beyond Dumping*. Westport, Conn.: Greenwood Press, 1984.

———, and Gary Davis. *America's Future in Toxic Waste Management: Lessons from Europe*. New York: Quorum Books, 1987.

Portney, Kent E. *Siting Hazardous Waste Treatment Facilities*. Westport, Conn.: Auburn House, 1991.

BOOK CHAPTERS AND JOURNAL ARTICLES

Andrews, Richard N.L. "Hazardous Waste Facility Siting: State Approaches." In Charles Davis and James P. Lester, eds., *Dimensions of Hazardous Waste Politics and Policy*. Westport, Conn.: Greenwood Press, 1988.

Applegate, Howard, and Richard Bath. "Hazardous and Toxic Substances in U.S.-Mexico Relations." *Texas Business Review* (September–October, 1983).

Arnott, Robert. "Waste Management in Northern Europe." *Waste Management and Research*, No. 4 (1985).

Barke, Richard. "Policy Learning and the Evolution of Federal Hazardous Waste Policy." *Policy Studies Journal* (September 1985).

Bowman, Ann O'M. "Intergovernmental and Intersectoral Tensions in Environmental Policy Implementation: The Case of Hazardous Waste." *Policy Studies Review* (November 1984).

———. "Hazardous Waste Management: An Emerging Policy Area Within an Emerging Federalism." *Publius* (Winter 1985).

———. "Hazardous Waste Cleanup and Superfund Implementation in the Southeast." *Policy Studies Journal* (September 1985).

———. "Explaining State Response to the Hazardous Waste Problem." *Hazardous Waste* (Fall 1984).

———. "Superfund Implementation: Five Years and How Many Cleanups?" In Davis and Lester, eds., *Dimensions of Hazardous Waste Politics and Policy*. Westport, Conn.: Greenwood Press, 1988.

———, and James Lester. "Hazardous Waste Management: State Government Activity or Passivity?" *State and Local Government Review* (Winter 1985).

Carnes, Sam A. "Confronting Complexity and Uncertainty: Implementation of Hazardous Waste Management Policy." In Dean E. Mann, ed., *Environmental Policy Implementation*. Lexington, Mass.: Lexington Books, 1982.

Cohen, Steven. "Defusing the Toxic Time Bomb: Federal Hazardous Waste Programs." In Norman Vig and Michael Kraft, eds., *Environmental Policy in the 1980s: Reagan's New Agenda*. Washington, D.C.: CQ Press, 1984.

———, and Marc Tipermas. "Superfund: Preimplementation Planning and Bureaucratic Politics." In James Lester and Ann Bowman, eds., *The Politics of Hazardous Waste Management*. Durham, N.C.: Duke University Press, 1983.

Davis, Charles E. "Approaches to the Regulation of Hazardous Wastes." *Environmental Law,* 18, No. 3 (1988).

———. "Implementing the Resource Conservation and Recovery Act of 1976." *Public Administration Quarterly* (Summer 1985).

———. "Perceptions of Hazardous Waste Policy Issues Among Public and Private Sector Administrators." *Western Political Quarterly* (September 1985).

———. "Public Involvement in Hazardous Waste Facility Siting Decisions." *Polity* (Winter 1987).

———. "Substance and Procedure in Hazardous Waste Facility Siting." *Journal of Environmental Systems*. No. 1 (1984–85).

———, and Richard Feiock. "Testing Theories of State Hazardous Waste Regulation." *American Politics Quarterly* 20 (October 1992).

———, and Joe Hagan. "Exporting Hazardous Waste: Issues and Policy Implications." *International Journal of Public Administration*, No. 4 (1986).

———, and James P. Lester. "Hazardous Waste and the Policy Process." In Davis and Lester, eds., *Dimensions of Hazardous Waste Politics and Policy*. Westport, Conn.: Greenwood Press, 1988.

Elliot, Michael. "Improving Community Acceptance of Hazardous Waste Facilities Through Alternative Systems for Mitigating and Managing Risk." *Hazardous Waste* (Fall 1984).

Feiock, Richard, and Charles Davis. "Can State Hazardous Waste Regulation Be Reconciled with Economic Development? An Empirical Assessment of the Goetze-Rowland Model." *American Politics Quarterly*, 19 (July 1991).

———, and C.K. Rowland. "Environmental Regulation and Economic Development: The Movement of Chemical Production Among States." *Western Political Quarterly*, 43 (September 1990).

Fitzgerald, Michael, Amy McCabe, and David Folz. "Federalism and the

Environment: The View from the States." *State and Local Government Review* (Fall 1988).

Getz, Malcolm, and Benjamin Walter. "Environmental Policy and Competitive Structure: Implications of the Hazardous Waste Management Program." *Policy Studies Journal*, 9 (Winter 1980).

Goetze, David. "A Decentralized Mechanism for Siting Hazardous Waste Disposal Facilities." *Public Choice*, 39 (1982).

———, and C.K. Rowland. "Explaining Hazardous Waste Regulation at the State Level." *Policy Studies Journal*, 14 (September 1985).

Goldfarb, William. "The Hazards of Our Hazardous Waste Policy." *Natural Resources Journal*, 19 (1979).

Gormley, William T., Jr. "Intergovernmental Conflict on Environmental Policy: The Attitudinal Connection." *Western Political Quarterly*, 40 (June 1987).

Gramling, George, and William Earl. "Cleaning Up After Federal and State Pollution Programs: Local Government Hazardous Waste Regulation." *Stetson Law Review*, 17 (1988).

Grunbaum, Werner. "Developing a Uniform Federal Common Law for Hazardous Waste Liability." *Policy Studies Journal* (September 1985).

Hadden, Susan G., Joan Veillette, and Thomas Brandt. "State Roles in Siting Hazardous Waste Disposal Facilities: From State Preemption to Local Veto." In Lester and Bowman, eds., *The Politics of Hazardous Waste Management*. Durham, N.C.: Duke University Press, 1983.

Hird, John A. "Superfund Expenditures and Cleanup Priorities." *Journal of Policy Analysis and Management*, 9 (Fall 1990).

Jorling, T.C. "Hazardous Substances in the Environment." *Ecology Law Quarterly*, 9 (1981).

Kamieniecki, Sheldon, Robert O'Brien, and Michael Clarke. "Environmental Policy and Aspects of Intergovernmental Relations." In David Morgan and Edwin Benton, eds., *Intergovernmental Relations and Public Policy*. Westport, Conn.: Greenwood Press, 1986.

Kraft, Michael, and Ruth Kraut. "The Impact of Citizen Participation on Hazardous Waste Policy Implementation: The Case of Clermont County, Ohio." *Policy Studies Journal* (September 1985).

Kramer, Kenneth W. "Institutional Fragmentation and Hazardous Waste Policy: The Case of Texas." In Lester and Bowman, eds., *The Politics of Hazardous Waste Management*. Durham, N.C.: Duke University Press, 1983.

Landy, Marc. "Ticking Time Bombs!!! EPA and the Formulation of Superfund." In Helen Ingram and Kenneth Godwin, eds., *Public Policy and the National Environment*. Greenwich, Conn.: JAI Press, 1985.

Lester, James P. "Hazardous Waste and Policy Implementation: The Subnational Role." *Hazardous Waste* (Fall 1985).

———. "The Process of Hazardous Waste Regulation: Severity, Complexity and Uncertainty." In Lester and Bowman, eds., *The Politics of Hazardous Waste Management*. Durham, N.C.: Duke University Press, 1983.

———. "Implementing Intergovernmental Regulatory Policy: The Case of Hazardous Waste." In Shyarnal Majumdar, E. Willard Miller, and Robert Schmalz, eds., *Management of Hazardous Wastes*. Phillipsburg, N.J.: Pennsylvania Academy of Science, 1989.

———. "New Federalism and Environmental Policy." *Publius* (Winter 1986).

———, James Franke, Ann O'M. Bowman, and Kenneth Kramer. "Hazardous Wastes, Politics, and Public Policy: A Comparative State Analysis." *Western Political Quarterly* (June 1983).

———, and Ann O'M. Bowman. "Subnational Hazardous Waste Policy Implementation: A Test of the Sabatier-Mazmanian Model." *Polity*, 21 (Summer 1989).

Lieber, Harvey. "Federalism and Hazardous Waste Policy." In Lester and Bowman, eds., *The Politics of Hazardous Waste Management*. Durham, N.C.: Duke University Press, 1983.

Mangun, William. "A Comparative Analysis of Hazardous Waste Policy Formulation Efforts Among West European Countries." *Policy Studies Journal* (September 1985).

Marshall, Patrick. "Not in My Backyard!" Editorial Research Reports. Washington, D.C.: Congressional Quarterly, Inc., June 9, 1989.

Matheny, Albert and Bruce Williams. "Knowledge vs. NIMBY: Assessing Florida's Strategy for Siting Hazardous Waste Disposal Facilities." *Policy Studies Journal* (September, 1985).

Mazmanian, Daniel, and David Morell. "The Elusive Pursuit of Toxics Management." *The Public Interest,* No. 90 (Winter 1988).

———. "The NIMBY Syndrome: Facility Siting and the Failure of Democratic Discourse." In Norman Vig and Michael Kraft, eds., *Environmental Policy in the 1990s*. Washington, D.C.: CQ Press, 1990.

Mitchell, Robert, and Richard Carson. "Property Rights, Protest, and the Siting of Hazardous Waste Facilities." *American Economic Review* (May 1986).

Morell, David L. "Technological Policies and Hazardous Waste Politics in California." In Lester and Bowman, eds., *The Politics of Hazardous Waste Management*. Durham, N.C.: Duke University Press, 1983.

Mumme, Stephen. "Dependency and Interdependence in Hazardous Waste Management Along the U.S.-Mexico Border." *Policy Studies Journal* (September 1985).

O'Brien, Robert, Michael Clarke, and Sheldon Kamieniecki. "Open and Closed Systems of Decision-Making: The Case of Toxic Waste Management." *Public Administration Review* (July/August 1984).

O'Hare, Michael. "Not on My Block You Don't: Facility Siting and the Importance of Compensation." *Public Policy* (Fall 1977).

O'Leary, Rosemary. "The Impact of Federal Court Decisions on the Policies and Administration of the U.S. EPA." *Administrative Law Review*, 41 (Fall 1989).

Portney, Kent. "Allaying the NIMBY Syndrome: The Potential for Compensation in Hazardous Waste Treatment Facility Siting." *Hazardous Waste* (Fall 1984).

———. "The Potential of the Theory of Compensation of Mitigating Public Opposition to Hazardous Waste Treatment Facility Siting: Some Evidence From Five Massachusetts Communities." *Policy Studies Journal* (September 1985).

Powell, John Duncan. "Assault on a Precious Commodity: The Local Struggle to Protect Groundwater." *Policy Studies Journal* (September 1985).

Riley, Richard. "Toxic Substances, Hazardous Wastes, and Public Policy:

Problems in Implementation." In Lester and Bowman, eds., *The Politics of Hazardous Waste Management*. Durham, N.C.: Duke University Press, 1983.

Ristoratore, Mario. "Siting Toxic Waste Disposal Facilities in Canada and the United States: Problems and Prospects." *Policy Studies Journal* (September 1985).

Rosenbaum, Walter. "The Politics of Public Participation in Hazardous Waste Management." In Lester and Bowman, eds., *The Politics of Hazardous Waste Management*. Durham, N.C.: Duke University Press, 1983.

Rowland, C.K., and Roger Marz. "Gresham's Law: The Regulatory Analogy." *Policy Studies Review* (November 1981).

———, S.C. Lee, and David Goetze. "Longitudinal and Catastrophic Models of Hazardous Waste Regulation." In Davis and Lester, eds., *Dimensions of Hazardous Waste Politics and Policy*. Westport, Conn.: Greenwood Press, 1988.

Schwartz, Seymour, Wendy Cockovich, Cecilia Fox, and Nancy Ostrom. "Improving Compliance with Hazardous Waste Regulations among Small Businesses." *Hazardous Waste and Hazardous Materials*, 6 (Summer 1989).

Senkan, Selim M., and Nancy W. Stauffer. "What to Do with Hazardous Waste." *Technology Review*, 84 (November/December 1981).

Sheehan, Michael. "Economism, Democracy, and Hazardous Wastes: Some Policy Considerations." In Sheldon Kamieniecki, Robert O'Brien, and Michael Clarke, eds., *Controversies in Environmental Policy*. Albany: SUNY Press, 1986.

Shue, Henry. "Exporting Hazards." *Ethics* (July 1981).

Toffner-Clausen, John. "Danish Hazardous Waste System." In John Lehman, ed., *Hazardous Waste Disposal*. New York: Plenum Press, 1983.

Walter, Benjamin and Malcolm Getz. "Social and Economic Effects of Toxic Waste Disposal." In Kamieniecki, O'Brien, and Clarke, eds., *Controversies in Environmental Policy*. Albany: SUNY Press, 1986.

Wells, Donald. "Site Control of Hazardous Waste Facilities." *Policy Studies Review* (May 1982).

Wiley, Karen, and Steven Rhodes. "Decontaminating Federal Facilities: The Case of the Rocky Mountain Arsenal." *Environment*, 29 (April 1987).

Williams, Bruce A., and Albert R. Matheny. "Hazardous Waste Policy in Florida: Is Regulation Possible?" In Lester and Bowman, eds., *The Politics of Hazardous Waste Management*. Durham, N.C.: Duke University Press, 1983.

———. "Testing Theories of Social Regulation: Hazardous Waste Regulation in the American States." *Journal of Politics* (May 1984).

Wolbeck, Bernd. "Political Dimensions and Implications of Hazardous Waste Disposal." In John Lehman, ed., *Hazardous Waste Disposal*. New York: Plenum Press, 1983.

Worthley, John A., and Richard Torkelson. "Managing the Toxic Waste Problem: Lessons from the Love Canal." *Administration and Society*, 13 (1981).

———. "Intergovernmental and Public-Private Sector Relations in Hazardous Waste Management: The New York Example." In Lester and Bowman,

The Politics of Hazardous Waste Management. Durham, N.C.: Duke University Press, 1983.

Wurth-Hough, Sandra. "Chemical Contamination and Governmental Policymaking: The North Carolina Experience." *State and Local Governmental Review* (May 1982).

Zimmerman, Rae. "Federal-State Hazardous Waste Management Policy Implementation in the Context of Risk Uncertainties." In Davis and Lester, eds., *Dimensions of Hazardous Waste Politics and Policy*. Westport, Conn.: Greenwood Press, 1988.

PUBLISHED REPORTS

Alm, Leslie R., and Gary Silverman. *Hazardous Wastes and Materials Guidebook*. Bowling Green, Ohio: Center for Governmental Research and Public Affairs, Bowling Green State University, 1991.

Association of State and Territorial Solid Waste Management Officials. *State Programs for Hazardous Waste Site Assessments and Remedial Actions*. Washington, D.C.: ASTSWMO, 1987.

Center for Hazardous Waste Management. *Coalition on Superfund Research Report*. Chicago: Illinois Institute of Technology, September 1989.

Commission for Economic Development, State of California. *Poisoning Prosperity: The Impact of Toxics on California's Economy*. Sacramento, Calif.: Commission for Economic Development, 1985.

Condon, Margaret, and Dixie Sipher. *Hazardous Waste Facility Siting: A National Survey*. Albany, N.Y.: Legislative Commission on Toxic Substances and Hazardous Wastes, June 1987.

Congressional Budget Office. *Hazardous Waste Management: Recent Changes and Policy Alternatives*. Washington, D.C.: Government Printing Office, 1985.

———. *Federal Liabilities Under Hazardous Waste Laws*. Washington, D.C.: Government Printing Office, May, 1990.

Conservation Foundation. *State of the Environment: 1984*. Washington, D.C.: Conservation Foundation, 1984.

———. *State of the Environment: A View Toward the Nineties*. Washington, D.C.: Conservation Foundation, 1987.

Council of State Governments. *Waste Management in the States*. Lexington, Ky.: The Council of State Governments, 1982.

Doyle, Paul. *Hazardous Waste Management: An Update*. Denver, Colo.: National Conference of State Legislatures, 1987.

Governor's Hazardous Waste Policy Advisory Council. *Hazardous Waste: A Management Perspective*. Tallahassee: The Institute of Science and Public Affairs, Florida State University, 1981.

ICF Incorporated. *Analysis of Community Involvement in Hazardous Waste Site Problems: A Report to the Office of Emergency and Remedial Response*. Washington, D.C.: U.S. Environmental Protection Agency, July 1981.

Kamlet, Kenneth S. *Toxic Substances Programs in U.S. States and Territories: How Well Do They Work?* Washington, D.C.: National Wildlife Federation. 1980.

Keystone Center. *State-Federal Relations in the RCRA Regulatory Program.* Keystone, Colo.: The Keystone Center, June, 1989.

Lennett, David, and Linda Greer. *State Regulation of Hazardous Waste: A Summary and Analysis.* Washington, D.C.: Environmental Defense Fund, 1985.

Mazmanian, Daniel, Michael Stanley-Jones, and Miriam Green. *Breaking Political Gridlock.* Claremont: California Institute of Public Affairs, 1988.

Minnesota Waste Management Board. *Charting a Course: Public Participation in the Siting of Hazardous Waste Facilities.* Crystal, Minn.: Waste Management Board, 1981.

National Conference of State Legislatures. *Hazardous Waste Management: A Survey of State Legislation, 1982.* Denver, Colo.: NCSL, 1982.

Organization for Economic Cooperation and Development. *Economic Aspects of International Chemicals Control.* Paris: OECD, 1983.

Paparian, Michael, Patricia Wells, and Peter Fearey. *Integrated Hazardous Waste Systems in the Federal Republic of Germany and Denmark.* Sacramento: California Foundation on the Environment and the Economy, 1984.

U.S. Congress. Office of Technology Assessment. *Technologies and Management Strategies for Hazardous Waste Control.* Washington, D.C.: Government Printing Office, 1983.

———. *Serious Reduction of Hazardous Waste.* Washington, D.C.: Government Printing Office, 1986.

———. *From Pollution to Prevention: A Progress Report on Waste Reduction.* Washington, D.C.: Government Printing Office, June 1987.

U.S. Environmental Protection Agency. *An Analysis of State Superfund Programs.* Washington, D.C.: Government Printing Office, September 1989.

———. *The Nation's Hazardous Waste Program at a Crossroads: The RCRA Implementation Study.* Washington, D.C.: Government Printing Office, July 1990.

U.S. General Accounting Office. *Hazardous Waste: Federal Civil Agencies Slow to Comply with Regulatory Requirements.* GAO/RCED-86-76, May 1986.

———. *Efforts to Clean Up DOD-Owned Inactive Hazardous Waste Sites.* GAO/NSIAD-85-41, April 12, 1985.

———. *Environmental Funding: DOE Needs to Better Identify Funds for Hazardous Waste Compliance.* GAO/RCED-88-62, December 1987.

———. *Hazardous Waste: EPA's Generation and Management Data Need Further Improvement.* GAO/PEMD-90-3, February 1990.

———. *Hazardous Waste: Issues Surrounding Insurance Availability.* GAO/RCED-88-2, October 1987.

———. *Hazardous Waste: New Approach Needed to Manage the Resource Conservation and Recovery Act.* GAO/RCED-88-115, July 1988.

———. *Hazardous Waste: Groundwater Conditions at Many Land Disposal Facilities Remain Uncertain.* GAO/RCED-88-29, February 1988.

———. *Hazardous Waste: Pollution Claims Experience of Property/Casualty Insurers.* GAO/RCED-91-59, February 1991.

———. *Hazardous Waste: The Cost and Availability of Pollution Insurance.* GAO/PEMD-89-6, October 1988.

———. *Inspection, Enforcement and Permitting Activities at the New Jersey and Tennessee Hazardous Waste Facilities*. GAO/RCED-84-7, June 1984.

———. *Superfund: A More Vigorous and Better Managed Enforcement Program Is Needed*. GAO/RCED-90-22, December 1989.

———. *Superfund: Interim Assessment of EPA's Enforcement Program*. GAO/RCED-89-40BR, October 1988.

———. *Hazardous Waste: Many Enforcement Actions Do Not Meet EPA Standards*. GAO/RCED-88-140, June 1988.

Wright, J. Ward. *Managing Hazardous Wastes: A Programmatic Approach*. Lexington, Ky.: Council of State Governments, 1986.

Index

**Note*: Refers to end-of-chapter notes.